SUSTAINABLE ETHANOL

Biofuels, Biorefineries, Cellulosic Biomass, Flex-Fuel Vehicles, and Sustainable Farming for Energy Independence

Jeffrey Goettemoeller
and
Adrian Goettemoeller

Prairie Oak Publishing
Maryville, Missouri

SUSTAINABLE ETHANOL
Biofuels, Biorefineries, Cellulosic Biomass, Flex-Fuel Vehicles, and
Sustainable Farming for Energy Independence

Copyright © 2007 by Jeffrey Goettemoeller and Adrian Goettemoeller

All rights reserved.

ISBN 978-0-9786293-0-4

Library of Congress Control Number: 2007929388
Library of Congress Subject Headings:
Alcohol as fuel–United States
Alcohol fuel industry–United States
Biomass energy

1.0

Prairie Oak Publishing
221 South Saunders St.
Maryville MO 64468
www.PrairieOakPublishing.com
www.ethanolbook.com

Contents

List of Figures — 6
Acknowledgements — 7
Disclaimer — 8
Introduction — 9

1. **A Brief History of Ethanol Fuel** — 11
   Lamp Fuel and the First Oil Well — 12
   Internal Combustion and the First Fuel Alcohol Movement — 12
   Chemurgy and the Second Fuel Alcohol Movement — 14
   Oil Embargoes and the Third Fuel Alcohol Movement — 16
   The Return of Cheap Oil — 17
   The End of Cheap Oil & the Rise of Ethanol — 18

2. **Will Cheap Oil Return?** — 23
   Easy Energy — 24
   The Rise and Fall of Big Oil — 24
   The Search for "Easy Oil" — 25
   Geopolitical Considerations — 26
   Growing World Oil Consumption — 27
   How Long will Oil Production Grow? — 28

3. **Economic and Security Benefits** — 31
   Economic Impact of Biorefineries — 31
   Tax Incentives for Ethanol and Oil — 32
   Farm Subsidies — 34
   Growing Oil Imports — 34
   The Hidden Cost of Imported Oil — 35
   Hurricane Vulnerability — 36
   Cellulosic Diversification — 37

4. **Environmental Impact** — 39
   Measuring Environmental Impact — 39
   Greenhouse Gases — 40
   Brazilian Ethanol — 42
   Groundwater Pollution — 42
   Air Pollution — 44
   Feedstock Sourcing — 46

5. **E10, E85, and Flex-Fuel Vehicles** — 51

   Running on E10 — 51
   E10 Price and Fuel Economy — 53
   E10 Availability — 54
   Running on E85 — 54
   Flex-Fuel Vehicles — 55
   Purchasing a Flex-Fuel Vehicle — 57
   E85 Availability — 58
   E85 Cost and Fuel Economy — 59

6. **Improving Fuel Economy on Ethanol** — 65

   Fuel Economy on E10 — 66
   Improving Flex-Fuel Vehicles — 68
   Ethanol and Hybrid Electric Vehicles — 72
   Ethanol Boosting with Direct Injection — 74
   Ethanol, Hydrogen, and Fuel Cells — 76
   Hydrated Ethanol — 76

7. **Food, Farming, and Land Use** — 83

   Food Prices — 84
   Food AND Fuel from Corn — 85
   Ethanol and World Hunger — 86
   Fossil Fuels and Agriculture — 88
   Sustainable Farming — 89
   Land Use Issues — 91
   Diversifying Energy Crops — 93
   High-Diversity Grassland — 95
   Enhancing Food Production with Energy Farming — 96

8. **Ethanol Production** — 99

   New Feedstocks — 101
   Ethanol from Sugar Feedstocks — 103
   Ethanol from Sweet Sorghum — 105
   Ethanol from Jerusalem Artichokes — 106
   Ethanol from Food Waste — 107
   On-Farm Ethanol from Waste Fruit — 108
   Ethanol from Beets — 108
   Ethanol from Corn Kernels — 109
   Adding Value to Coproducts — 112
   Ethanol from Field Peas — 115
   Ethanol from Grain Sorghum, Wheat, and Barley — 115
   Reducing Process Fuel Use — 116

Reducing Water Use — 118
Alternative Process Fuels — 119
Combined Heat and Power — 122
Ethanol-Livestock Integration — 124
Small-Scale Ethanol Production — 126
Ethanol Transportation & Pipeline Issues — 127
Butanol: The Other Alcohol — 128

9. **Cellulosic Ethanol** — 137
   Commercializing Cellulosic Production — 138
   Cellulosic Conversion Technologies — 143
   Biochemical Methods — 144
   Thermochemical Methods — 145
   Biogas as a Transportation Fuel — 146
   Waste & Coproduct Feedstocks — 148
   Agricultural Residue Feedstocks — 149
   Dedicated Cellulosic Energy Crops — 151
   A Sticky Coproduct — 148
   Harvest and Transportation of Feedstocks — 155
   The Pyrolysis Route to Cellulosic Ethanol — 155
   Regional Biomass Processing Centers — 156
   Pipeline Transportation of Corn Stover Silage — 157
   Economics of Cellulosic Ethanol — 158
   How Much Ethanol Can We Make? — 158

10. **Energy Balance: Is Ethanol Renewable?** — 165
    All BTU's are Not the Same — 165
    Fossil Energy Replacement Ratio — 166
    Petroleum Replacement Ratio — 167
    Rating Cellulosic Ethanol — 168
    Comparing Ethanol and Gasoline — 169
    Fuel Economy and Energy Balance — 170
    Variables and Trends — 171

11. **Facing our Energy Future** — 175
    Balancing our Energy Budget — 176

Ethanol Questions and Answers — 178
Selected Resources/Bibliography — 181
Glossary — 184
Index — 188

Figures

1-1.	Fuel Ethanol Production, 1982–2006 —	**19**
2-1.	U.S. Crude Oil Spot Price, 1995–2007 —	**23**
2-2.	Percentage of Proved World Oil Reserves by Region, 2005 —	**27**
2-3.	World Oil Consumption, 1985–2005 —	**28**
3-1.	Estimated Revenue Loss from Tax Incentives for Petroleum and Ethanol —	**33**
5-1.	Same location U.S. National Average Retail Gasoline and E85 Prices —	**60**
5-2.	Break-even Prices for Gasoline Alternatives Based on Fuel Economy Reduction —	**62**
6-1.	Change in Fuel Economy Using E10 Relative to Ethanol-free Gasoline —	**67**
7-1.	Average Price Paid U.S. Farmers per Bushel of Corn, 1980-2007 —	**83**
7-2.	Corn Planted Since 1992 per Market Year —	**92**
8-1.	Ethanol Production Paths —	**100**
8-2.	Ethanol Production by Feedstock, 2006 —	**101**
8-3.	Estimated Ethanol Yield by Feedstock —	**102**
8-4.	Estimated Production Cost by Feedstock —	**104**
8-5.	Ethanol Production from Corn Kernels —	**111**
8-6.	Conventional vs. Combined Heat & Power —	**124**
9-1.	The Biochemical Cellulosic Production Process —	**145**
9-2.	Annual Biomass Potential from U.S. Forests and Agriculture —	**160**
10-1.	Calculating Fossil Energy Replacement Ratio for Corn Ethanol —	**166**
10-2.	Calculating Petroleum Replacement Ratio for Corn Ethanol —	**168**
10-3.	Energy Balance Ratios for Ethanol and Gasoline —	**169**
10-4.	U.S. Natural Gas Wellhead Price, 1996–2007 —	**172**

Acknowledgements

We extend our thanks and admiration to the farmers, entrepreneurs, politicians, journalists, investors, scientists, engineers, and others striving to make ethanol production and use more efficient and sustainable. This book is about the technologies making ethanol make sense, yet these technologies are reflections of the dreams, ingenuity, foresight, and hard work of many people, past and present.

Thanks to Ed Malewski for composing a poem for this book. Thanks also to those who provided comments, corrections, testimonials, and encouragement. Thanks especially to friends and family members. Their support and advice is indispensable.

Disclaimer

This book is designed to provide accurate information about the subject matter covered and is sold with the understanding that the publisher and authors are not providing legal, accounting, investment guidance, or other professional services. This book is not intended to provide all the information on the subject covered or all the information available to the authors and/or publisher. Internet addresses, trade, firm, corporation, or company names, product names, and other resources are provided for informational purposes only and do not constitute endorsement by the authors or publisher.

While every effort has been made to assure the accuracy of this book, there may be mistakes in content and typography. Therefore, this book should be used as a general introduction to the subject matter covered and not as the ultimate source of information on ethanol, biofuels, and other subjects covered.

This book is written for informational and entertainment purposes only. The authors and Prairie Oak Publishing accept neither liability nor responsibility to any person or entity with respect to loss, risk, or damage caused or alleged to be caused, directly or indirectly, by information in this book.

If you do not want to be bound by the disclaimer printed above, you may return this book to the publisher for a full refund.

Introduction

The era of cheap oil is over and America is increasingly dependent on imported energy. In our search for energy security and sustainability, is ethanol a viable option or a dead end? Is it truly renewable? Is it good for the environment? In short, is ethanol good for America? This book goes beyond the headlines in search of answers to America's ethanol questions.

Most North American automobiles run on gasoline, and this will not change quickly. We need ethanol in the near term because it can be used with gasoline in existing vehicles. Soon, new technologies could improve the fuel economy of ethanol-powered vehicles. Ethanol could also be used in combination with electric motors, fuel cells, turbo-chargers, and hydrogen technology.

Critics of ethanol often implicate current farming methods. Today's ethanol is mostly made from corn grown in an unsustainable manner, with large amounts of petroleum-derived fertilizers and pesticides. However, a growing number of farmers are transitioning to more sustainable practices. Higher fossil fuel prices will accelerate this trend. Farmers are getting better yields with fewer fossil fuel inputs and less soil erosion.

Ethanol producers are using fewer fossil fuel inputs as well. Some are even using renewable "process fuels" in place of natural gas or coal. Before long, they will be making cellulosic ethanol, butanol, and biogas from prairie grasses, crop residues, and various organic waste materials.

The way we make and use ethanol can be improved dramatically. We are just scratching the surface of ethanol's potential. Ethanol is not a perfect fuel. There is no such thing. But ethanol is preferable to imported petroleum, and ethanol production is getting more efficient. This book is about the technologies making ethanol make sense. These technologies and the people behind them are why we believe ethanol is good for America.

Chapter 1

A Brief History of Ethanol

In his 2006 State of The Union Address, U.S. President George W. Bush said America is "addicted to oil."[1] Is our society ready to make the sacrifices necessary to break the addiction? Since the beginning of the petroleum era, ethanol and other biofuels have been waiting in the wings. Time after time, we have chosen cheap oil over renewable alternatives.

The history of ethanol fuel is an epic struggle between ethanol and petroleum, played out in a series of circumstances—taxes, wars, discoveries, prohibition, inventions, and champions on each side.[2] In market-driven economies, cost usually determines dominant energy sources. One way to reduce our dependence on foreign oil is to make domestically produced alternatives available at a competitive price. This looked possible when world events caused oil prices to soar. But these were temporary interruptions in a flood of cheap oil, a bounty formed over countless centuries. This flood always inundated alternative energy sources. Each emergence of a fuel ethanol industry, however, left us with valuable lessons in preparation for the inevitable end of cheap fossil fuels.

In this chapter, we will just scratch the surface of this fascinating story. The comprehensive history of fuel alcohol is contained in *The Forbidden Fuel: Power Alcohol in the Twentieth Century*, by Hal Bernton, William Kovarik, and Scott Sklar, originally published in 1982. An updated edition is in production.

Lamp Fuel and the First Oil Well

The growing use of ethanol fuel in our time completes a circle begun long before the automobile era. In the early 1800's, back when ethanol was simply referred to as "alcohol," it was an important lamp fuel ingredient. The popularity of alcohol as beverage and fuel made it a convenient target for taxation. A $2.00 per gallon excise tax was applied to alcohol in 1862 to help pay for the Civil War. This made it too expensive for illumination purposes.[3] As if this crushing tax wasn't enough, the first commercial oil wells had already ushered in a less expensive alternative to alcohol.[4]

If policy makers of the 1800's had known America would one day be held hostage over oil, they might have tilted the tables in favor of ethanol rather than fossil fuels. Instead, America embarked on an addiction to oil that now threatens our security, economy, and environment.

Internal Combustion and the First Fuel Alcohol Movement

While alcohol fuel was stymied by U.S. taxes in the middle of the 19th century, the same was not true in Europe. In 1860, German engineer Nikolas Otto used alcohol as the fuel for one of his "Otto Cycle" combustion engines.[5] Toward the end of the 1800's, alcohol-fueled engines enjoyed a period of ascendancy in Europe as a response to worries over the long-term stability of petroleum supplies.[6]

Back in North America, farmers were beginning to exploit the deep, rich soil of the Great Plains. Production spikes caused commodity prices to fall, a situation that would often plague America's farmers. Despite the steep tax on alcohol, Henry Ford designed his first car, the Quadricycle, to run on pure ethanol in the 1896.[7] The majority of Americans were farmers at the turn of the century, giving them strong political influence. With the help of President Theodore Roosevelt, the Civil

War era alcohol tax was finally ended in 1906.[8] The way was clear for America's first farmer-driven fuel alcohol movement.

In 1908, Henry Ford equipped his "Model T" with engines capable of running on ethanol, gasoline, or a combination of the two.[9] The ability to use multiple fuels would later be termed "flex-fuel." Other manufacturers made provisions for alcohol fuel as well. Despite the best efforts of power alcohol enthusiasts, however, alcohol was never able to seriously challenge less expensive gasoline as the dominant transportation fuel.

Around 1917 to 1918, Word War I triggered a surge in demand for industrial alcohol. Production reached 50–60 million gallons per year, marking the high point of America's first fuel alcohol movement.[10] After the war, proponents continued to preach the advantages of alcohol fuel, but demand dropped as inexpensive fossil fuels once again dominated the marketplace.

Around this time, ethanol missed out on a huge opportunity. Automotive engineers were looking for gasoline additives that would prevent engine knock at higher compression ratios for better fuel economy. Ethyl alcohol (today referred to as ethanol), already well known as a knock inhibitor, was a possible choice. It was discovered, however, that tetraethyl lead could also do the job. Radford University professor Bill Kovarik discovered documentation from the 1920's indicating tetraethyl lead was originally intended as a stopgap solution until ethyl alcohol production could be ramped up. Tetraethyl lead became the dominant anti-knock additive for decades. Eventually, tetraethyl lead was recognized as a serious health hazard. In 1986, it was finally banned as a fuel additive.[11]

After World War I, prohibition dashed hopes for maintaining a robust fuel alcohol industry. Beginning in 1920, the 18th amendment to the U.S. Constitution prohibited the "manufacture, sale, or transportation of intoxicating liquors within, the importation thereof into, or the exportation thereof from the United States and all territory subject to the jurisdiction thereof for beverage purposes."[12] It was still legal to manufacture and use alcohol for fuel, but doing so must have attracted close scrutiny from authorities. Also, fuel alcohol had to be mixed with petroleum so it could not be used as a beverage.[13] This requirement is still in effect today. Prohibition was eventually repealed by the ratification of the 21st amendment in 1933, marking a new period of opportunity for fuel alcohol.[14]

Chemurgy and the Second Fuel Alcohol Movement

A 1930's movement known as chemurgy was based on the idea of using agricultural products for industrial purposes. Henry Ford continued to advocate fuel alcohol during prohibition. Meanwhile, chemist William J. Hale popularized the principles of chemurgy through enthusiastic books, articles, and speeches. He believed farmers and the economy would benefit from new markets for surplus farm products. In 1936, he released a book entitled *Prosperity Beckons: Dawn of the Alcohol Era*, preaching the benefits of an alcohol fuel industry.[15]

Hale predicted economic disaster for farmers in the wake of World War I. His prediction came true. American farmers enjoyed relative prosperity during the war as demand from Europe drove up prices for farm products. The war prevented European farmers from producing as they once had. Meanwhile, American farmers were successful at ramping up yields with the use of new machinery, fertilizers, and other new technologies. When the war ended and demand for farm commodities plummeted, so did the prices farmers could ask for their goods. The onset of the Great Depression in the 1930's made matters even worse as few people had the means to purchase farm products.

Crude oil was inexpensive during the 1930's, but so were farm commodities. It made sense for farmers to embrace the chemurgy movement as they fought for economic survival. They needed better prices for their crops.

In 1937, leaders in the chemurgy movement established a large-scale production facility for fuel alcohol in Atchison, Kansas. The Atchison Agrol Company marketed low-level ethanol blends through independent oil distributors and farm cooperatives. Over 2000 service stations, mostly in the Midwest, were offering "agrol" by the spring of 1938.[16] But the good times for agrol were short-lived. Writing in 1939, William J. Hale described the downfall of the Atchison Agrol Company:

> ...when appreciable quantities of agrol motor fuel found distribution at any one point, we found at this exact point a barrage of denunciation by antiquated gasoline supporters of all that pertained to agrol. No matter how silly the remarks, the effect of this continued criticism reduced sales to prospective customers. ...No new industry can long withstand such unfair competition, even though it be financed entirely by philanthropists. Thus, after a year's successful operation, though not altogether under efficient management, it has seemed best to discontinue the manufacture of agricrude alcohol till that time when Government or humanitarian agencies can take hold of this greatest of all industries and fight for the nation's good against un-American interests now in control.[17]

Opposition from supporters of gasoline may have hastened the demise of the Atchison Agrol Company, but competition from cheap oil also played a role. In addition, grain prices improved as the nation emerged from the Great Depression. Hale, ever the optimist, focused on lessons learned:

> The Atchison plant has fulfilled its mission. American farmers are given a true picture of what this mighty industry holds in store for them and their unending prosperity. Fermentation industries are afforded data that will contribute to greater efficiency in operation.[18]

The onset of World War II temporarily revived the alcohol industry once again. Alcohol was needed for synthetic rubber production, vital for the war effort. In 1942, Japan captured the principle rubber producing regions of the Far East, where 90% of the world's natural rubber production was located.[19] Initially, government contracts for synthetic rubber went to companies using petroleum as the feedstock. But these companies were unable to ramp up production quickly enough. Alcohol-based synthetic rubber was called upon to fill the void. Whiskey distilleries were converted to industrial alcohol production, new ethanol biorefineries were built, and the old Atchison agrol biorefinery was put back in use.[20] Over 600 million gallons per year of new alcohol production capacity was created.[21]

Eventually, petroleum-based synthetic rubber was perfected and expanded. Natural rubber also became available again. Grain prices began to rise due to growing demand from Europe after the war. The alcohol biorefineries built for the war effort were sold to private spirit distillers or allowed to sit idle and deteriorate. By 1949, over 90% of U.S. industrial alcohol production used natural gas as the principle feedstock.[22] Meanwhile, gasoline use soared as post-war Americans took to the highways in ever-growing numbers.

Oil Embargoes and the Third Fuel Alcohol Movement

In 1973, America learned a difficult lesson about reliance on imported petroleum. In October of that year, Egypt and Syria attacked Israel, sparking the Yom Kippur war. Oil exporting Arab nations slashed production by 5 million barrels per day and stopped exports to the United States and other supporters of Israel.[23] American motorists were faced with gas shortages and a dramatic increase in prices at the pump—from 38.5 cents per gallon in May 1973 to 55.1 cents per gallon in June 1974.[24] An Iranian regime change in 1979 produced another disruption in oil markets and a 120% increase in oil prices in one year's time.[25] While these disruptions were temporary, the dangers of reliance on foreign oil had become a reality. The stage was set for America's third fuel alcohol movement.

In the 1970's, farmers grew increasingly dependent on fossil fuel-derived inputs such as fertilizers, pesticides, herbicides, and tractor fuel. Soaring fossil fuel prices drove up expenses. A few farmers decided to grow their own fuel. They began to produce ethanol in small-scale biorefineries, bringing about the beginnings of today's thriving ethanol industry.

By 1980, the federal government enacted new regulations making small-scale ethanol production easier. For facilities with a capacity less than 10,000 gallons per year, the requirements for bonding and inspection were removed.[26] By the end of 1980, over 200 farm-based ethanol biorefineries were under construction.[27]

In the 1970's and 1980's, ethanol was marketed in the form of "gasohol," a blend of up to 10 percent ethanol with gasoline. In 1978, the first gasohol pump opened for business in Lincoln, Nebraska. With the help of a four-cent per gallon tax exemption, gasohol was available at over 10,000 service stations in fifty states by the fall of 1981.[28] This rapid expansion prompted the emergence of a larger-scale ethanol industry dominated by grain processing companies.

The journey from that first gasohol pump in 1978 to wide acceptance in the new millennium was not easy for the ethanol industry. From the mid 1980's to the late 1990's, an old foe threatened again—cheap oil.

The Return of Cheap Oil

The price spikes of the 1970's helped bring about a frenzied worldwide search for oil. Oil companies directed exploration efforts toward non-OPEC nations thought to be safer for investment—most notably Alaska, Mexico, and the North Sea. Meanwhile, oil consumption was falling thanks to conservation measures and a worldwide economic decline. The world went from a state of oil shortage to surplus in just a few years. U.S. dependence on foreign oil went from 46.5% in 1977 to 27% in 1985.[29] OPEC began losing market share, prompting additional price cuts. While prices fluctuated in following years thanks to events such as Iraq's invasion of Kuwait, oil was a bargain in historical terms well into the 1990's.

As during past times of low oil prices, it was difficult for ethanol to compete as a primary transportation fuel. Many producers were forced out of business. The number of commercial ethanol biorefineries in the United States peaked at 163 in 1984. By the end of 1985, only 74 of these facilities remained in operation.[30]

The ethanol industry survived through the 1980's and 1990's, thanks in large part to ethanol's usefulness as an oxygenate additive. Added to gasoline at 5–10%, ethanol causes a cleaner burn and fewer emissions. Reducing emissions became important in large cities with increasing numbers of automobiles. The first use of ethanol as an oxygenate was in Denver, Colorado in 1988 for winter reduction of carbon monoxide emissions.[31] The clean air act amendments of 1992 required the use of oxygenate additives in certain cities. The primary oxygenate at the time was MTBE (Methyl Tertiary Butyl Ether, made from natural gas and petroleum), but ethanol was used in some regions.[32] The third fuel alcohol movement had stalled, but the ethanol industry survived, ready to rebound in the new millennium.

The End of Cheap Oil & the Rise of Ethanol

Demand for ethanol began to grow after MTBE was discovered in groundwater. Beginning in 1999, some states limited its use in motor fuel.[33] By 2006, nearly 20 states had banned the use of MTBE in gasoline.[34] This assured a steady market for ethanol, the primary replacement for MTBE. A 2005 decision by federal lawmakers refusing legal protection with regard to MTBE use hastened the rise of ethanol.[35]

The first decade of the new millennium witnessed another dramatic rise in energy prices. Unlike the brief price spikes of the 1970's, this upward trend would last for years and continues today. During the second week of February 1998, the average world crude oil price was $9.31 per barrel. From that time, prices rose steadily. Peaks were higher, and dips failed to go as low (figure 2-1). By the first week of June 2006, crude oil went for $64.67 per barrel.[36] Ethanol was again competitive with gasoline. It made sense to blend low levels of ethanol with gasoline even where it was not mandated. Some states mandated the use of 10% ethanol in gasoline, citing benefits for farmers, economies, national security, and the environment.

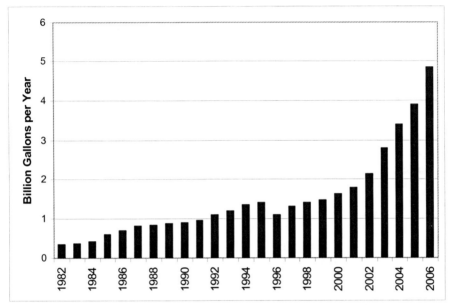

Figure 1-1: Fuel Ethanol Production, 1982–2006 (Data from the Renewable Fuels Association, www.ethanolrfa.org)

Eventually, some of the world's major oil companies jumped on board the biofuels bandwagon.[37] Ethanol production rose over 150% from 2000 to 2005.[38] In order to maintain this kind of growth, the industry needs investment in new technologies such as cellulosic feedstocks. Past investors in alternative energy were often burned by the return of cheap oil. Are the factors causing high oil prices likely to go away soon, or will oil prices remain high for an extended period of time? Let's explore this question in Chapter 2.

Notes

1. White House, "President Bush Delivers State of the Union Address," news release, January 31, 2006, http://www.whitehouse.gov/news/releases/2006/01/20060131-10.html.
2. For those of you interested in a more mythical summary of the ethanol story, the following poem was submitted by an environmentalist friend:

   The Redemption of Alcohol
   by Ed Malewski
   *Come to find that fossil fuel is the stuff that feeds the flames of Hell,
   and the Devil dealt a deal of death with those who had their souls to sell
   that cars would crave His gasoline and thereby make them well-to-do
   while alcohol, the runner-up, was wrongly called a devil's brew.
   Thus, Mankind lost its feeble mind on Mai Tais, whiskey sours, and gin,
   the air got choked on CO_2 and Hell filled up with wealthy men.
   But Mankind sobered up at last to see the Devil's evil mess
   so the only need for oil became to heal cracked feet, anoint the blessed,
   and moonshine took its rightful place in tanks of our American Dream,
   then global warming slowly cooled. . . the flowers bloomed, and Heaven beamed.*

3. U.S. Energy Information Administration, "Ethanol Timeline," *Energy Kid's Page*, http://www.eia.doe.gov/kids/history/timelines/ethanol.html.
4. U.S. Energy Information Administration, *Energy in the United States: 1635–2000*, http://www.eia.doe.gov/emeu/aer/eh/petro.html.
5. U.S. Energy Information Administration, "Ethanol Timeline."
6. William Kovarik, "Henry Ford, Charles F. Kettering and the Fuel of the Future," *Automotive History Review* 32 (Spring 1998): 7–27. Reproduced on the Web at http://www.radford.edu/~wkovarik/papers/fuel.html.
7. U.S. Energy Information Administration, "Ethanol Timeline."
8. Kovarik, "Henry Ford, Charles F. Kettering."
9. U.S. Energy Information Administration, "Ethanol Timeline."
10. Ibid.
11. Kovarik, "Henry Ford, Charles F. Kettering."
12. The Constitution of The United States, *The National Archives*, http://www.archives.gov/national-archives-experience/charters/constitution_amendments_11-27.html.
13. U.S. Energy Information Administration, "Ethanol History," *Energy Kid's Page*, http://www.eia.doe.gov/kids/energyfacts/sources/renewable/ethanol.html.
14. The Constitution of The United States, *The National Archives*.
15. William J. Hale, *Prosperity Beckons: Dawn of the Alcohol Era* (Boston: The Stratford Company, 1936).

16. Hal Bernton, William Kovarik, and Scott Sklar, *The Forbidden Fuel: Power Alcohol in the Twentieth Century* (New York: Boyd Griffin, 1982), 24.
17. William J. Hale, *Farmward March* (New York: Coward-McCann, 1939), 132–133.
18. Ibid., 133.
19. Paul Wendt, "The Control of Rubber in World War II," *The Southern Economic Journal* 13 (January, 1947): 203.
20. Bernton, Kovarik, and Sklar, *The Forbidden Fuel*, 28–29.
21. Kovarik, "Henry Ford, Charles F. Kettering"
22. Bernton, Kovarik, and Sklar, *The Forbidden Fuel*, 33.
23. Stuart E. Eizenstat, Testimony before the U.S. Senate Committee on Commerce, Science, and Transportation on the National Security Implications of Increased CAFE Standards, January 24, 2002, http://commerce.senate.gov/hearings/012402eizenstat2.pdf.
24. Edward J. Markey, "Economics of Dependence on Foreign Oil - Rising Gasoline Prices," Opening Statement at a hearing before the U.S. House of Representatives Select Committee on Energy Independence and Climate Change, May 9, 2007, http://globalwarming.house.gov/pdf/050907EJMstatement.pdf.
25. Stuart E. Eizenstat, Testimony before the U.S. Senate Committee.
26. Bernton, Kovarik, and Sklar, *The Forbidden Fuel*, 44.
27. Ibid., 55.
28. Ibid., 59.
29. Edward J. Markey, "Economics of Dependence on Foreign Oil."
30. U.S. Energy Information Administration, "Ethanol Timeline."
31. Ibid.
32. Ibid.
33. Ibid.
34. Timothy B. Wheeler, "Refiners to Phase out Use of MTBE," *Baltimore Sun*, February 17, 2006.
35. Juliet Eilperin, "Protection for Fuel Additive Dropped," *Washington Post*, July 27, 2005.
36. U.S. Energy Information Administration.
37. du Pont de Nemours and Company, "DuPont, BP Announce Partnership to Develop Advanced Biofuels," news release, June 6, 2006; Chevron Corporation, "Chevron Pursues Opportunities in Emerging Biofuels Sector," news release, May 31, 2006, http://www.chevron.com/news/press/2006/2006-05-31.asp.
38. Based on data from the Renewable Fuels Association, www.ethanolrfa.org/industry/statistics/#A.

Chapter 2

WILL CHEAP OIL RETURN?

Oil price declines of the 1980's put a damper on many renewable energy initiatives. Through the late 1980's and 1990's, oil was so cheap that ethanol and other biofuels found little room for growth. Oil has been more expensive recently, but will this last?

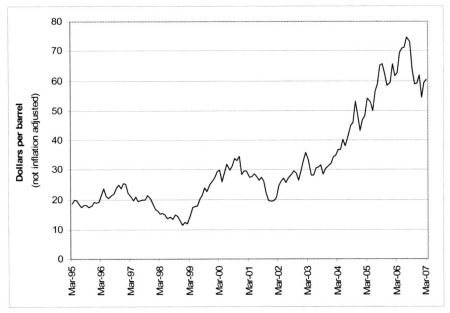

Figure 2-1: U.S. Crude Oil Spot Price, 1995–2007 (WTI Cushing OK FOB. Data from U.S. Energy Information Administration)

We believe a long-term collapse in oil prices is unlikely this century. Trillions of barrels of oil remain to be tapped, but these barrels will be more difficult to access and therefore more expensive. This should allow sustainable biofuels to compete more successfully. Shorter duration declines in oil prices are likely, however, and renewable energy industries need to be prepared for them. In addition, government policies should be designed to help renewable energy survive slumps in energy prices.

Easy Energy

The era of cheap oil was made possible by solar energy, photosynthesis, geologic processes, and time. The energy potential in living matter, originally spread through huge spans of time and space, was concentrated deep under the earth's surface in the form of fossil fuels. This concentration of energy, resulting in a high energy density, is what makes petroleum so useful.

With ideal conditions, the energy required to extract crude oil is minimal in comparison to its energy potential. In other words, oil can have a very good energy returned on energy invested (EROEI). The earth itself has done the heavy lifting for us. But the earth works at a slow pace. We are cashing in on an investment made over millions of years. The problem is, easily exploited oil reserves are steadily declining. On average, the extraction of remaining oil reserves will result in a less favorable EROEI.

With the production of biofuels, we attempt to densify organic biomass more quickly. Through the application of modern technology in cooperation with natural processes, it is possible to optimize the sustainable densification of plant-derived energy into biofuels suitable for existing automobiles.

The Rise and Fall of Big Oil

The first successful U.S. oil well was drilled in 1859 at Oil Creek, Pennsylvania.[1] Oil had already been discovered and produced near Baku, Azerbaijan in Asia.[2] The development of the internal combustion engine in the 1870's and 1880's created a vast and growing market for liquid fuels. After a brief

TWO | WILL CHEAP OIL RETURN?

period of ascendancy for alcohol fuel, this market was largely captured by fossil fuels. Seven petroleum companies eventually dominated the world oil trade.[3] They found and produced huge quantities of light sweet crude. But today, production from many of the giant oil fields is declining.

Through the past century, some have predicted the end of cheap oil only to be proven wrong by subsequent events. What is different this time? Trends pointing to the possible end of cheap oil include:

1. Not enough high quality, easily extracted oil reserves are available.
2. Geopolitical conflicts prevent access to many of the world's best oil reserves.
3. Increasing world oil consumption could outpace supplies.

The Search for "Easy Oil"

According to the U.S. Energy Information Administration, only 63% of world crude oil and natural gas liquids production was replaced by new reserves from 2003 to 2005.[4] The area of the former Soviet Union and Eastern Europe was the only region where new reserves exceeded production.

Petroleum varies in quality. The vast majority used over the past century has been relatively light and sweet—easily refined into gasoline and other products. Recently, oil companies have turned to sour crudes high in sulfur as well as heavy, thick crudes that in some cases, such as the Canadian tar sands, do not flow until heated. The fact that companies are beginning to tap these unconventional reserves shows that conventional reserves are insufficient. Market forces compel companies to produce oil at the least possible expense. That is why lower quality, harder to produce oil is generally left in place until easily accessible reserves run low.

A 2000 study by the U.S. Geological Survey identified possible areas of new oil production, but most are in areas with high production costs or environmental concerns. Untapped reserves are thought to exist in the high Arctic, the coasts of Greenland, and the ultra-deep waters off the coasts of North America, South America, Africa, and Asia.

Geopolitical Considerations

Governments of some oil-producing countries are imposing higher taxes or even nationalizing production of oil and natural gas. The largest remaining reserves of light and heavy crude oil are off-limits to foreign oil companies in many places.[5] Privately held oil companies have small shares of world oil production and reserves, limiting their influence on world oil prices.[6] Oil production of western majors has been flat, with little prospect for improvement. According to the U.S. Federal Trade Commission, the share of world crude oil production accounted for by U.S.-based companies declined from 11.4% in 1990 to 8.4% in 2002.[7]

Some observers, including oil investor Matthew Simmons, believe oil reserves in the Middle East may be smaller than we realize.[8] This is a serious concern because a majority of the world's known reserves are in the Middle East (figure 2-2).

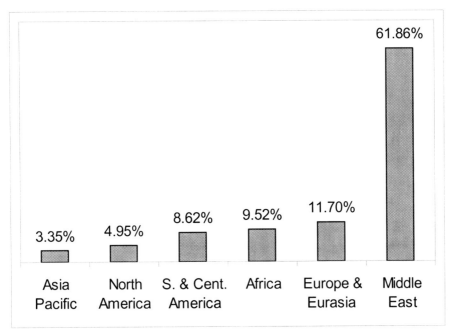

Figure 2-2: Percentage of Proved World Oil Reserves by Region, 2005 (Data from the *BP Statistical Review of World Energy 2006*)

Growing World Oil Consumption

Any growth in oil consumption is likely to put upward pressure on prices. The U.S. Energy Information Administration (EIA) expects worldwide oil consumption to increase from about 80 million barrels per day in 2003 to almost 120 million barrels per day in 2030.[9] China and other growing Asian economies will likely be responsible for the much of the increase. Since 1985, oil consumption grew faster in the Asia Pacific region than in North America (figure 2-3). Indeed, Asia Pacific oil use had almost caught up to North American consumption by 2005.

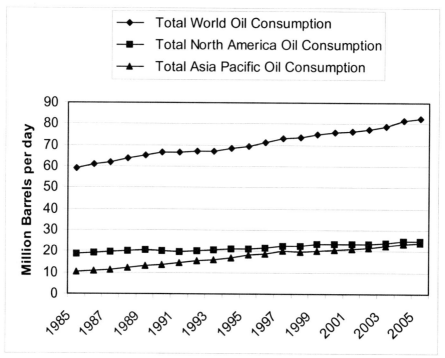

Figure 2-3: World Oil Consumption, 1985–2005 (Data from the *BP Statistical Review of World Energy 2006*)

How Long will Oil Production Grow?

Future oil prices depend in large part on our ability to increase world production from the current level of around 82 million barrels per day. Some, including Matthew Simmons, believe oil production may have already peaked.[10] Even assuming more optimistic projections, we need to cut oil use and develop alternatives as soon as possible. The United States Geological Survey projects 2037 as the year in which oil production will peak (year 2000 report).[11] This is a short time when faced with a major change in world energy systems. It took many decades and much economic pressure for the world to convert from coal to petroleum-based fuels. Breaking our addiction to oil might take even longer.

In a 2005 report, the United States Army Corp of Engineers issued this warning:

> US production of all types of oil and natural gas are past their peaks. World-wide consumption of energy is forecast to increase 60% from current levels by 2030 and future availability of customary energy sources is problematic. Growing domestic (US) consumption will continue to increase dependence on foreign and potentially unstable energy sources. Current energy policies and consumption practices are not sustainable. Oil production is approaching its peak; low growth in availability can be expected for the next five to ten years.[12]

When the United States Army concludes world oil production is approaching its peak, perhaps it is time to take notice. The U.S. Government Accountability Office agrees with the need for action. In a 2007 report to Congress, they highlighted the difficulty in forecasting oil use and production, and the potential dangers of energy shortages. The title of their report summarizes their findings:

> *CRUDE OIL: Uncertainty about Future Oil Supply Makes It Important to Develop a Strategy for Addressing a Peak and Decline in Oil Production.*[13]

Notes

1. R.C. Selly, *Elements of Petroleum Geology, 2nd Edition* (San Diego: Academic Press, 1998), 3.
2. Ibid, 416.
3. Ibid, 3.
4. U.S. Energy Information Administration, *Performance Profiles of Major Energy Producers 2005,* http://www.eia.doe.gov/emeu/perfpro/tab08.htm.
5. J. Robinson West, Testimony for the U.S. Senate Commerce Committee, September 21, 2005.
6. U.S. Federal Trade Commission, "FTC Issues Staff Report on 'The Petroleum Industry: Mergers, Structural Change, and Antitrust Enforcement'," news release, August 13, 2004, http://www.ftc.gov/opa/2004/08/oilmergersrpt.shtm.
7. Ibid.

8. Matthew Simmons, *Twilight in the Desert: The Coming Saudi Oil Shock and the World Economy*, (Hoboken: John Wiley and Sons, 2005).
9. U.S. Energy Information Administration's International Energy Outlook 2006.
10. Scott Malone, "World oil production may have peaked," *Reuters.com*, October 26, 2006, http://today.rueters.com/.
11. USGS World Energy Assessment Team, "U.S. Geological Survey World Petroleum Assessment 2000," *U.S. Department of the Interior U.S. Geological Survey*, 2000.
12. Donald F. Fournier and Eileen T. Westervelt, "Energy Trends and Their Implications for US Army Installations" *United States Army,* September 2005.
13. United States Government Accountability Office, *CRUDE OIL: Uncertainty about Future Oil Supply Makes It Important to Develop a Strategy for Addressing a Peak and Decline in Oil Production,* February 2007, http://www.gao.gov/new.items/d07283.pdf.

Chapter 3

ECONOMIC AND SECURITY BENEFITS

For too long, the price farmers received for their crops did not even cover the cost of production. Until recently, grain prices remained flat while the cost of other goods steadily rose. Thanks in part to biofuel production, grain prices are finally reaching the point where farmers and rural economies can survive without crop subsidies.

Economic Impact of Biorefineries

Many farmers are now part owners of ethanol biorefineries, bringing jobs and investment to rural areas. A study completed for the Renewable Fuels Association by John M. Urbanchuk, Director LECG LLC, quantifies benefits to the U.S. economy from the U.S. ethanol industry in a single year—2006:[1]

- $23.1 billion added to the nation's Gross Domestic Product
- 163,034 new jobs
- $6.7 billion added to household income
- $2.7 billion in tax revenue for the federal government
- $2.2 billion in tax revenue for state and local governments
- Nearly 5 billion gallons of ethanol production reduced the need for oil imports by 206 million barrels.

According to Urbanchuk, the benefit to a local economy from an ethanol biorefinery is up to 40% greater with local ownership compared to absentee-owned facilities.[2] He notes considerable growth in local ownership since the early 1990's.[3] A 2006 Iowa State University study found a typical Iowa ethanol biorefinery with no local ownership creates 133 jobs in the regional economy. That figure grows by 29 jobs for every 25% increase in local ownership.[4]

Recent explosive growth in ethanol production is reversing the trend toward local ownership. Outside investors want a piece of the action. To some extent, this is necessary for such a rapid expansion. But farmer and local ownership is still desirable. The Institute for Local Self Reliance proposes steps for maximizing ethanol's benefit to rural economies. They include educational efforts, mechanisms allowing farmer-owners of ethanol facilities to extract equity while maintaining continued local ownership, and federal ethanol incentives that favor farmer and local ownership.[5] The full report, *Ownership Matters: Three Steps to Ensure a Biofuels Industry that Truly Benefits Rural America*, is available at:

www.ilsr.org/

Tax Incentives for Ethanol and Oil

The ethanol industry benefits from government subsidies in the form of tax incentives. This is not unusual for industries deemed important by society. When evaluating anything to do with ethanol, it makes sense to do so in comparison to the commodity it replaces. In the case of ethanol, this would be crude oil—the feedstock for gasoline. Even though the oil industry is more established and mature, we still subsidize it in various ways for the sake of national and economic security. We should do no less for biofuels that can be used in place of oil.

It is difficult to compare government subsidies for the ethanol and oil industries because of the difference in maturity and size. The government-produced data in figure 3-1 at least make the point that both industries have received considerable direct assistance from tax incentives. Ethanol is not unique in receiving government assistance. It is just one of many commodities we have decided to support as a nation for the common good.

Tax incentive	Summed over years	Dollars in millions Adjusted to year 2000 dollars
Petroleum industry		
Excess of percentage over cost depletion[a]	1968–2000	$81,679–$82,085
Expensing of exploration and development costs[a]	1968–2000	42,855–54,580
Alternative (nonconventional) fuel production credit	1980–2000	8,411–10,542
Oil and gas exception from passive loss limitation	1988–2000	1,065[b]
Credit for enhanced oil recovery costs	1994–2000	482–1,002
Expensing of tertiary injectants	1980–2000	330[c]
Ethanol industry		
Partial exemption from the excise tax for alcohol fuels	1979–2000	7,523–11,183
Income tax credits for alcohol fuels	1980–2000	198–478

Note: inflation-adjusted summations of estimated revenue losses for petroleum and ethanol fuel tax incentives from 1968 to 2000. Data developed from unadjusted annual revenue loss estimates made by the Department of the Treasury and the staff of the Joint Committee on Taxation (JCT). When two figures are provided for an incentive, they represent the estimates developed from Treasury's and JCT's data.

a/ In some years, revenue losses associated with other fuels and non-fuel minerals were included with revenue losses from oil and gas.
b/ There is no JCT revenue estimate because only Treasury recognizes this tax code provision as a separate tax incentive.
c/ There is no Treasury revenue estimate because only JCT recognizes this tax code provision as a separate tax incentive.

Figure 3-1: Estimated Revenue Loss from Tax Incentives for Petroleum and Ethanol (Source: United States General Accounting Office, September 25, 2000)

Farm Subsidies

With the rising price of corn in recent years, ethanol tax incentives are offset by lower farm subsidy payments. Under terms of a farm bill in effect from 2002 to 2007, Counter-cyclical payments for U.S. corn crops have gone from over 905 million dollars per year in 2005 to nothing in 2007 (USDA forecast).[6] Counter-cyclical payments are based on market prices. The price of corn is finally above the level that triggers counter-cyclical payments, thanks in large part to demand from ethanol producers.

Growing Oil Imports

In 2006, 59.6% of all oil consumed in the United States came from other countries, up from 34.8% in 1973.[7] The top five exporters of petroleum to the United States are:[8]

1. Canada
2. Mexico
3. Saudi Arabia
4. Venezuela
5. Nigeria.

Canada has large, but dwindling conventional oil reserves. It has been investing heavily in new heavy oil production from bitumen (tar) sands, but this oil is expensive to produce, as it must be mined and then upgraded from bitumen to conventional oil. Saudi Arabia is reported to have the world's largest conventional oil reserves, but its moderate government is coming under increasing pressure from Saudi citizens who reject any cooperation with the United States or Europe. If the Saudi government falls, its oil exports may be sharply curtailed in the ensuing chaos. Venezuela has large deposits of both conventional and heavy oil, but its leaders have become increasingly hostile to the United States, raising questions about the reliability of its oil exports.

Oil shipping is vulnerable as well. A large number of oil tankers pass through a handful of narrow "chokepoints" that could be blocked by terrorist groups or unfriendly governments.[9] The Strait of Hormuz—the entrance to the Persian Gulf—is of particular concern because of the large oil reserves in the Persian Gulf region.

During the Iranian-prompted oil shocks of the 1970's, oil supply interruptions and panic caused actual fuel shortages in North America. It wasn't just a matter of rising prices—fuel was simply not available for a time. With our growing reliance on imported oil, supply interruptions would be more dangerous than ever. The portion of U.S. petroleum imports coming from the Persian Gulf region, for instance, rose from 13.6% in 1973 to 16.2% in 2006.[10]

In 2006 testimony before the U.S. Senate Foreign Relations Committee, Milton Copulos, president of the National Defense Council Foundation, warned of grave consequences should our oil supply be interrupted again:

> The supply disruptions of the 1970s cost the U.S. economy between $2.3 Trillion and $2.5 Trillion. Today, such an event could carry a price tag as high as $8 Trillion—a figure equal to 62.5 percent of our annual GDP or nearly $27,000 for every man, woman and child living in America.
>
> But there is more cause for concern over such an event than just the economic toll. A supply disruption of significant magnitude, such as would occur should Saudi supplies be interdicted, would also dramatically undermine the nation's ability to defend itself.
>
> Oil has long been a vital military commodity, but today has taken on even more critical importance.[11]

The Hidden Cost of Imported Oil

For the United States, the National Defense Council Foundation estimates the hidden cost of imported oil at 825.1 billion dollars *in just one year* (2006).[12] This amounts to around $2700.00 per year for every person. Their study considered costs related to defense expenditures, loss of current economic activity due to capital outflow, loss of domestic investment, loss of government revenues, and the cost of periodic oil supply disruptions.

The hidden costs of imported petroleum make ethanol look like a bargain when you realize little petroleum goes into the production of ethanol. Most of the fossil fuels used in corn ethanol production are sourced from within North America.

Hurricane Vulnerability

With the concentration of infrastructure along the gulf coast, North America's energy supply is vulnerable to hurricanes. In 2004 and 2005, hurricanes shut down a large portion of U.S. oil and natural gas production and refining. The following is testimony by the U.S. Department of Energy on hurricane damage during 2005:

> Hurricane Katrina struck the Gulf Coast on August 29, several days after first landing in south Florida. It left an unprecedented amount of destruction in an area totaling 90,000 square miles... Eleven petroleum refineries were shut down, representing 2.5 million barrels per day—or nearly one-sixth—of U.S. refining capacity. With Katrina, more than a quarter of U.S. crude oil production—1.4 million barrels per day—was shut in. Nearly 9 billion cubic feet per day of natural gas production in the federal Gulf of Mexico was shut in, representing 17 percent of U.S. gas production, with additional production losses occurring in areas under Louisiana's jurisdiction. The Louisiana Offshore Oil Port (LOOP) was shut down, as were a number of major oil and gas pipelines. As a consequence, pipeline deliveries of gasoline, diesel, jet fuel, and propane supplies to the east coast and southeastern states were halted.
>
> Hurricane Rita made landfall on September 24, and did even greater harm to our nation's energy markets than Katrina. After Hurricane Rita, 19 refineries were shut down, representing nearly a third of U.S. refining capacity. In the federal Gulf of Mexico, virtually all crude production and eighty percent of natural gas production was shut in. 27 natural gas processing facili-

ties were shuttered—representing half of Gulf Coast natural gas processing capability. Offshore rigs and platforms suffered damage. The LOOP (Louisiana Offshore Oil Port) was shut down once again, along with a number of major pipelines.[13]

While lessons learned from 2005 will mitigate the impact of future hurricanes, concentration of infrastructure still poses significant risk. Domestic ethanol production can help reduce this vulnerability. Ethanol biorefineries are mainly located in the interior of the United States, away from hurricane danger.

Cellulosic Diversification

Growth of cellulosic ethanol production will prompt construction of biorefineries in more diverse locations, reducing distance to market. Cellulosic materials are available in every region of North America. Crop residue, wood waste, food waste, and municipal solid waste could be sourced, refined into ethanol, and consumed within a relatively limited region. Localized production of cellulosic ethanol could help accelerate decentralization of the U.S. energy infrastructure.

Notes

1. John M. Urbanchuk, *Contributions of the Ethanol Industry to the Economy of the United States*, Prepared for the Renewable Fuels Association by LECG LLC, February 2007, http://www.ethanolrfa.org/.
2. John M. Urbanchuk, "Economic Impacts on the Farm Community of Cooperative Ownership of Ethanol Production," *LECG LLC*, February 7, 2007, http://www.usda.gov/oce/forum/2007%20Speeches/PDF%20speeches/J%20Urbanchuk.pdf
3. Ibid.
4. Leopold Center for Sustainable Agriculture, "ISU Study Determines Regional Economic Values of Ethanol Production in Iowa," Iowa State University news release, September 22, 2006, http://www.leopold.iastate.edu/news/newsreleases/2006/ethanol_092206.htm.
5. David Morris, "Ownership Matters: Three Steps to Ensure a Biofuels Industry that Truly Benefits Rural America," *Institute for Local Self-Reliance*, April 2006, http://www.newrules.org/agri/ownershipbiofuels.pdf.

6. USDA Economic Research Service, *Farm Income and Costs: 2007 Farm Sector Income Forecast*, February 14, 2007, http://www.ers.usda.gov/; USDA, "Direct & Counter-cyclical Program," *Farm Bill Forum Comment Summary & Background*, http://www.usda.gov/.
7. U.S. DOE Energy Information Administration, "Overview of U.S. Petroleum Trade," *Monthly Energy Review March 2007*, http://www.eia.doe.gov/.
8. U.S. DOE Energy Information Administration, *Crude Oil and Total Petroleum Imports Top 15 Countries January 2007 Import Highlights,* April 2, 2007, http://www.eia.doe.gov/.
9. U.S. DOE Energy Information Administration, *World Oil Transit Chokepoints General Background*, http://www.eia.doe.gov/.
10. DOE Energy Information Administration, "Overview of U.S. Petroleum Trade," *Monthly Energy Review March 2007*, http://www.eia.doe.gov/.
11. Milton R. Copulos, *Testimony before the Senate Foreign Relations Committee*, March 30, 2006, http://foreign.senate.gov/.
12. The National Defense Council Foundation, *The Hidden Cost of Oil: An Update,* January 8, 2007, http://www.ndcf.org/.
13. U.S. Department of Energy Office of Public affairs, "Remarks Prepared for Energy Secretary Bodman," *Senate Energy and Natural Resources Committee Hearing,* October 27, 2005.

Chapter 4

ENVIRONMENTAL IMPACT

An energy system is unsustainable if it destroys our ecosystem. Experts often disagree about environmental impacts, perhaps because they are looking at different aspects of the issue. An examination of a broad range of issues shows that ethanol fuel is neither the one perfect solution (there is no such thing) nor a dead end.

Measuring Environmental Impact

Practically any human activity impacts the environment in some way. Making and using ethanol is no exception. Ethanol can be considered beneficial if it does less damage than the fuel it replaces. Ethanol primarily replaces petroleum-derived gasoline in today's North American transportation system. When we consider ethanol's environmental impact, we must do so in comparison to gasoline. We need to begin with some assumptions established elsewhere in this book:

1. Over time, fossil fuel prices will probably remain high (see Chapter 2).

2. Ethanol enjoys a better net energy balance than gasoline (see Chapter 10).

High fossil fuel prices will affect the way we grow and refine ethanol feedstocks for the better. The increasing cost of fossil fuel-based herbicides, fertilizers, and process fuels creates a market force to minimize their use. The resulting improvement in energy balance and decrease in harmful chemical use will lessen environmental damage from ethanol.

Most scientists believe we derive more energy value from the use of ethanol than could be derived from the fossil fuels used in its production. Of thirteen major studies on the subject completed between 1998 and 2005, nine showed a positive energy balance for corn ethanol.[1] Ethanol may perform better than the energy balance studies indicate. They generally assume a lower fuel economy for ethanol based on its lower energy content. However, some car models already on North American roads actually suffer little or no loss in fuel economy on E10 (10% ethanol).[2]

The numbers look even better for ethanol's energy balance when compared to gasoline. Gasoline has a decidedly negative energy balance because of the fossil fuels required for extracting, transporting, and refining crude oil. The University of Chicago's Argonne National Laboratory estimates gasoline has a negative 25% energy balance, while corn ethanol enjoys a positive energy balance of more than 25%.[3]

Ethanol's energy balance has steadily improved over the years.[4] This trend means less fossil fuel use and less environmental impact related to emissions, chemical contamination, and global warming. On the other hand, technology for refining gasoline is older and less likely to show as much improvement in efficiency.

Greenhouse Gases

While scientists disagree about the role of human activity in climate change, many attribute increases in global temperatures to a "greenhouse effect" caused by certain gases, some of which are emitted by internal combustion engines. Plants to be used as ethanol feedstocks extract carbon dioxide from the air as they grow. Carbon is taken out of the atmosphere and trapped in the form of plant material, offsetting greenhouse gas emissions from burning ethanol as fuel. The Argonne National Laboratory examined seventeen studies on greenhouse gas emissions resulting from using corn ethanol at 85–100% proportions. Thirteen of these studies showed a decrease in greenhouse gas emissions compared to ethanol-free gasoline.[5]

Today, the way we grow corn is ethanol's weak point. But this is getting better. No-till farming, precision farming, and new corn cultivars have already improved ethanol's energy balance (see Chapter 7). On-farm research in sustainable agriculture since the 1980's has proven that grain crops such as corn can be grown with greatly reduced fossil fuel inputs, while maintaining yield, reducing soil erosion, and building soil fertility.[6] Even greater improvements will accompany cellulosic ethanol and butanol production. Argonne National Laboratory estimates corn ethanol use achieves an 18–29% reduction in greenhouse gas emissions compared to gasoline, while cellulosic ethanol will achieve an 85–86% reduction.[7]

Results might be even better for cellulosic ethanol if perennial crops make up a large part of the feedstock. Perennials thrive on land unsuitable for grain crops. The roots of perennial crops sequester (store) more carbon than is contained in the aboveground portion of the plant. In the case of switchgrass, roots account for more than 80% of plant biomass.[8] Perennial root systems also produce ideal conditions for beneficial microorganisms and earthworms. According to a detailed 2006 analysis by the USDA Agricultural Research Service (ARS), perennial crops will reduce atmospheric greenhouse gases much more as compared to annual row crops.

> Compared with the life cycle of gasoline and diesel, ethanol and biodiesel from corn rotations reduced greenhouse gas emissions by 35–40%; reed canarygrass reduced emissions by 85%; and switchgrass and hybrid poplar reduced emissions by more than 115%.[9]

Any improvement in the fuel economy of ethanol-burning vehicles will also reduce greenhouse gas emissions, as less will be needed in order to go the same distance down the road. At least some of the fuel economy technologies featured in Chapter 6 are likely to be implemented as fossil fuels become more expensive.

Brazilian Ethanol

Brazil is home to a thriving ethanol industry, thanks in part to advantages of climate and a long history of technological and infrastructure development. Brazil is also famous for its rain forests. It is all too easy to jump to the conclusion that rain forests are being replaced by sugar cane production for ethanol. This notion is a myth. Brazil's sugar cane regions are far from the Amazon rain forests. The rain forested region is too hot and humid for optimum sugar cane growth.[10] Furthermore, according to Brazilian sugar cane grower Milton Maciel, Brazil's ethanol is made from around 3 million hectares (7.4 million acres) of sugar cane. This is only about 1% of Brazil's arable land. Brazil has increased total ethanol output mainly by boosting production per acre. According to Maciel, Brazil's ethanol yields have gone from 375 gallons/acre/year in 1975 to 870 gallons/acre/year in 2006. He notes yields from organic sugar cane are even higher.[11] Growth in yield/acre/year will continue. Ethanol industry executive Dr. Fernando Reinach calculates a future theoretical yield of over 2,300 gallons/acre/year based on new hybrid sugar cane varieties and improved technology.[12]

Groundwater Pollution

Early in the history of the automobile, octane-boosting substances were added to gasoline to suppress engine knock, thereby permitting higher compression ratios and better fuel economy.[13] Ethanol was one of the first recognized octane boosters, but tetraethyl lead was used instead, resulting in dangerous lead contamination.[14] The list of possible maladies resulting from lead exposure is long and ugly, including anemia, nervous system damage, brain damage, kidney damage, and even death.[15] The use of tetraethyl lead in gasoline was eventually banned by the EPA and completely phased out by 1996.[16]

Tetraethyl lead was replaced by other octane boosters that also proved to be dangerous for human health. One of these is a compound made from natural gas and petroleum known as methyl tertiary-butyl ether (MTBE).[17] Like ethanol, MTBE is an oxygenate. Oxygenates help gasoline burn more completely, reducing harmful smog-forming pollutants.[18] According to the EPA, however, MTBE is a potential human carcinogen at high levels:

> EPA's Office of Water has concluded that available data are not adequate to estimate potential health risks of MTBE at low exposure levels in drinking water but that the data support the conclusion that MTBE is a potential human carcinogen at high doses.[19] (U.S. Environmental Protection Agency)

Scientists are concerned about MTBE because it travels faster and persists longer underground compared to other gasoline components.[20] According to the EPA, the first report of major MTBE contamination was in Santa Monica, California in 1996. Half of the city's water supply was found to be heavily contaminated and two well fields were shut down.[21] Nearly 20 states had banned MTBE in motor fuel as of 2006.[22]

Ethanol can replace MTBE as an octane-boosting oxygenate. Other options exist, but they have their own environmental consequences. One alternative is to increase the proportion of aromatics resulting from the refining process.[23] But this increases gasoline's toxicity. Benzene, for instance, is a known carcinogen resulting from this process. It produces "temporary nervous system disorders, immune system depression, anemia."[24]

Ethanol, on the other hand, is much less toxic than MTBE or benzene. If ethanol was our only fuel and no other contaminates were already in the soil, it would be fairly harmless in itself. There are concerns, however, that ethanol spills may cause contaminant plumes of pre-existing aromatic hydrocarbon contaminants such as benzene to travel a greater distance and do somewhat more damage than they would without ethanol present.[25] This effect is known as cosolvency. A 2004 Brazilian study found cosolvency becomes a problem only with large spills.[26] The ultimate solution would be to clean up or remove contaminants that are the real source of the problem.

Ethanol's benefits appear to outweigh its risks when it comes to groundwater contamination. A 2000 study by the California Environmental Policy Council concluded ethanol use is preferable to continued use of MTBE.[27] If replacement of toxic octane boosters and oxygenates were the only environmental benefit of ethanol, it would be well worth using.

Air Pollution

Gasoline is a brew of chemicals causing complex interactions and emissions. As an oxygenate, ethanol improves fuel combustion properties and decreases harmful tailpipe emissions including carbon monoxide, hydrocarbons, and benzene.[28] This applies to low or high concentrations of ethanol.

Not all the effects are positive, though. Low-level blends such as E10 can create problems not associated with higher-level blends like E85. Low levels of ethanol tend to increase the volatility of fuel, resulting in greater evaporative emissions of harmful volatile organic compounds from the fossil fuel portion of the mixture, especially during warm weather. These emissions originate from the fuel tank or fuel lines. Refiners can compensate by formulating a lower volatility blendstock (base gasoline) during warmer months.[29] However, a complication results from co-mingling of the blendstock-adjusted E10 with standard gasoline. This co-mingling happens when drivers switch between E10 and ethanol-free gasoline during warm weather. Such mixing mitigates the effect of the low-volatility blendstock, resulting in higher volatility.[30]

The volatility issue may be solved by the on-board vapor recovery systems introduced in 1998. These systems trap up to 96% of fuel tank vapors.[31] Another possible solution involves blending butanol, a different type of alcohol, along with the ethanol. Butanol can mitigate vapor pressure problems caused by ethanol.[32] Read more about butanol in Chapter 8.

Today's higher compression vehicles require a certain level of octane for knock suppression. We are not going back to the very low compression, low fuel economy vehicles of the past. Seen in this light, ethanol is clearly our best option at this time. Its drawbacks are minor compared to those of other octane enhancers. A year 2000 report from the Institute for Local Self Reliance concludes that ethanol's air quality benefits, including reduction in emissions of particulate matter, outweigh its drawbacks.[33]

In the future, automobile technologies designed to take better advantage of ethanol are likely to cause a drastic decrease in air pollution (see Chapter 6). Due to the chemical properties involved, higher concentrations of ethanol such as E85 (85% ethanol) do not bring about the volatility problems associated with lower proportions of ethanol,[34] and many of the dangerous compounds are displaced by the less toxic ethanol. A Report from the U.S. Congressional Research Service points out the net benefits of ethanol in high-level blends:

> The air quality benefits from purer forms of ethanol can also be substantial. Compared to gasoline, use of E85 and E95 can result in a 30–50% reduction in ozone-forming emissions. And while the use of ethanol also leads to increased emissions of acetaldehyde, a toxic air pollutant, as defined by the Clean Air Act, these emissions can be controlled through the use of advanced catalytic converters.[35]

A life cycle analysis coordinated by the U.S. Department of Energy looked at E85 made with ethanol derived from corn stover (crop residue). They found corn stover ethanol reduces fossil energy use by 102% and greenhouse gas emissions by 113%. Results were mixed for impacts on air quality. CO, NO_x, and SO_x show increases, while hydrocarbon ozone precursors are lessened.[36]

The ethanol injection technology discussed in Chapter 6 could eliminate concerns about volatility by avoiding the need to mix ethanol and gasoline in the fuel tank. It could also boost fuel economy significantly. Extraction of hydrogen from ethanol for use in fuel cells could also be a step forward in lowering air pollution.

Feedstock Sourcing

Most North American ethanol is made from the starch portion of corn kernels. The efficiency and sustainability of corn farming has improved in recent years. Farmers still use significant quantities of petroleum-based fertilizers, herbicides, and tractor fuel, however, often resulting in environmental damage. Cultivation of perennial crops for cellulosic ethanol will be much more environmentally friendly. Perennial cropping nearly eliminates the need for pesticides, fertilizers, and yearly tillage.

CARBON SEQUESTRATION: Perennial grasses can sequester carbon (lowering atmospheric greenhouse gases) and yield ethanol at the same time. USDA Agricultural Research Service scientist Mark Liebig found that Switchgrass fields held 7 tons more carbon per acre than nearby corn and wheat fields!

> Greater soil carbon under switchgrass was observed at all depths, but it was most pronounced at one to three feet down—a depth in the soil profile where switchgrass has more root biomass than corn or wheat. Switchgrass roots grow as long as eight feet, compared to three to six feet for corn and wheat.[37] (USDA ARS)

The sites studied by Liebig are representative of about 74 million acres of the Northern Plains and northern Corn Belt. Evaluations of switchgrass are being conducted to determine if this deep storage of soil carbon holds true elsewhere.

On the gasoline side of the ledger, crude oil is the feedstock. Crude oil sourcing is unlikely to become significantly more environmentally friendly. In fact, growing scarcity of readily accessible oil reserves could prompt increased drilling in environmentally sensitive locations. Advances in oil tanker and pipeline technology might reduce oil spills, but improvements in this mature industry would be limited.

Environmentally speaking, no energy solution is perfect, but ethanol is an improvement over gasoline and an important part of a sustainable future. Ethanol's impact on the environment depends in large part on how it is produced and how it is used. On the use side, fuel-efficient vehicles optimized for ethanol would make ethanol more sustainable. On the production side, the high cost of fossil fuels is likely to encourage a trend toward sustainable farming methods, less fossil fuel input, and greater efficiency. Farmers and producers will need to play a leading role.

Past efforts at regulating environmental responsibility have generally been crushed by low oil prices. For the first time in the petroleum era, oil prices are likely to remain high enough to allow alternatives to take root.

Notes

1. Michael Wang, "The Debate on Energy and Greenhouse Gas Emissions Impacts of Fuel Ethanol" *University of Chicago Argonne National Laboratory,* 2005, 24.
2. KT Knapp, FD Stump, & SB Tejada, "The Effect of Fuel on the Emissions of Vehicles over a Wide Range of Temperatures," *Journal of the Air & Waste Management Association* 43 (July 1998); American Coalition for Ethanol, *Fuel Economy Study,* 2005, http://www.ethanol.org/documents/ACEFuelEconomyStudy.pdf.
3. Wang, "The Debate on Energy," 23.
4. Ibid., 24.
5. Ibid., 27.
6. Visit the Practical Farmers of Iowa web site for information about their long-term on-farm research in low-input farming methods: http://www.pfi.iastate.edu/ofr/Practices_and_Research.htm
7. Wang, "The Debate on Energy," 21. Greenhouse gas emission reductions per gallon of ethanol to displace an energy-equivalent amount of gasoline.
8. Mikkel Pates, "Researchers Unlocking Switchgrass Secrets," *CASMGS Insider,* http://www.casmgs.colostate.edu/insider/vigview.asp?action=2&titleid=511#.
9. USDA Agricultural Research Service, *Biodiversity Management in Northeastern Grazing Lands 2006 Annual Report,* 2006, http://ars.usda.gov/research/projects/

projects.htm?ACCN_NO=406621&showpars=true&fy=2006.
10. Milton Maciel, "Ethanol from sugar cane in Brazil," *Biofuels Now,* Oct 2006, http://www.biofuelsnow.com/.
11. Milton Maciel, "Ethanol from Brazil and the USA," *ASPO-USA /Energy Bulletin,* Oct 2006, http://www.energybulletin.net/21064.html.
12. Fernando Reinach, "Biofuels in Brazil: Today and in the Future," Presentation at the *4th annual Life Sciences & Society Symposium* at the University of Missouri, Columbia, March 15, 2007.
13. U.S. Environmental Protection Agency, *Clean Alternative Fuels: Ethanol,* 2002.
14. Jamie Lincoln Kitman, "The Secret History of Lead: Special Report," *The Nation,* March 20, 2000, http://www.thenation.com/doc/20000320/kitman/4.
15. U.S. Environmental Protection Agency, *Lead Compounds,* 2000, http://www.epa.gov/ttn/atw/hlthef/lead.html.
16. Ibid.
17. U.S. Energy Information Administration, *Ethanol Timeline,* http://www.eia.doe.gov/kids/history/timelines/ethanol.html.
18. U.S. Environmental Protection Agency, *Methyl Tertiary Butyl Ether (MTBE),* 2006, http://www.epa.gov/mtbe/gas.htm.
19. U.S. Environmental Protection Agency, *Methyl Tertiary Butyl Ether (MTBE),* 2006, http://www.epa.gov/mtbe/water.htm.
20. Michael J. Moran, John S. Zogorski, and Paul J. Squillace, "Occurrence and Implications of Methyl tert-Butyl Ether and Gasoline Hydrocarbons in Ground Water and Source Water in the United States and in Drinking Water in 12 Northeast and Mid-Atlantic States, 1993–2002," *U.S. Geological Survey,* 2004, 121.
21. U.S. Environmental Protection Agency, *Methyl Tertiary Butyl Ether (MTBE),* 2006, http://www.epa.gov/mtbe/water.htm.
22. Timothy B. Wheeler, "Refiners to Phase out Use of MTBE," *Baltimore Sun,* February 17, 2006.
23. U.S. Environmental Protection Agency, *Methyl Tertiary Butyl Ether (MTBE).*
24. U.S. Environmental Protection Agency, *Consumer Factsheet on: BENZENE,* 2006, http://www.epa.gov/safewater/dwh/c-voc/benzene.html.
25. Susan E. Powers, Pedro J. Alvarez, and David W. Rice, Chapter 1 in *Subsurface Fate and Transport of Gasoline Containing Ethanol* (University of California Lawrence Livermore National Laboratory, 2001), 5.
26. HX Corseuil, BI Kaipper, & M. Fernandes, *Water Research* 38 (March 2004): 1449–1456.
27. Ibid., 3.
28. Marika Tatsutani, ed., *Health, Environmental, and Economic Impacts of Adding Ethanol to Gasoline in the Northeast States, Volume 2, Air Quality, Health, and Economic Impacts* (Boston: Northeast States for Coordinated Air Use Management, 2001), 10.
29. Ibid., 13

30. Ibid., 14
31. Ibid.
32. Dupont, *Biobutanol FAQ,* 2006, http://www2.dupont.com/Biofuels/en_US/FAQ.html.
33. David Morris and Jack Brondum, "Does Ethanol Use Result in More Air Pollution?," *Institute for Local Self Reliance,* 2000.
34. Tatsutani, ed., *Health, Environmental, and Economic Impacts,* 13.
35. Brent D. Yacobucci and Jasper Womach, "RL30369: Fuel Ethanol: Background and Public Policy Issues" *Congressional Research Service,* 2000.
36. John Sheehan et al, "Energy and Environmental Aspects of Using Corn Stover for Fuel Ethanol," *Journal of Industrial Ecology,* 7 (Summer/Fall 2003): 117–146.
37. Don Comis, "Energy Farming with Switchgrass Saves Carbon," *USDA Agricultural Research Service,* July 19, 2006, http://www.ars.usda.gov/is/pr/2006/060719.htm.

Chapter 5

E10, E85, AND FLEX-FUEL VEHICLES

You might already be using ethanol in your car. E10 (10% ethanol) is widely available and approved by automobile makers. You might even be driving a vehicle that can burn up to 85% ethanol (E85). E85 is formulated for flex-fuel vehicles. These vehicles look the same as other cars, trucks, and vans, but run perfectly on anywhere from 0% to 85% ethanol.

Ethanol, also known as ethyl alcohol or fuel alcohol, is a potent version of drinking alcohol with the molecular formula CH_3CH_2OH. The federal government requires fuel alcohol to be denatured by adding up to 5% gasoline or other denaturant, "poisoning" the brew to prevent human consumption.[1]

The "E" in E10 and E85 stands for ethanol, while the numbers indicate ethanol percentage. E10 is 10% denatured ethanol and 90% gasoline. E85 is 85% denatured ethanol and 15% gasoline. Because the ethanol is denatured, E85 and E10 actually have slightly less than 85% or 10% ethanol content. Ethanol percentage also varies on a seasonal basis. E85 can contain as little as 75% denatured ethanol during the winter to help with cold weather starting.[2]

Running on E10

E10 is used in existing gasoline-powered engines. Ethanol is widely used as an oxygenate gasoline additive in areas with air pollution problems. Oxygen in ethanol makes fuel burn more cleanly. All major automobile makers warrant their vehicles to run on E10.[3] Since 1981, the U.S. has consumed more than 170 billion gallons of ethanol blends. Assuming a 20 miles per gallon average, this translates to nearly 3.5 trillion miles driven on ethanol blends.[4] The U.S. Environmental Protection Agency (EPA) lists several benefits of ethanol blends:

- Ethanol vehicles exhibit the same power, acceleration, payload, and cruise speed as conventionally fueled vehicles.

- Ethanol absorbs moisture and helps prevent gas-line freeze-up in cold weather, preventing the need to add expensive and possibly harmful fuel additives.

- Ethanol has some detergent properties that reduce buildup, which keeps engines running smoothly and fuel injection systems clean for better performance.[5]

The U.S. Department of Energy also endorses ethanol: "Blends of up to 10% ethanol with gasoline (E10) are approved for use in all gasoline vehicles, and have been used for many years across the nation to improve air quality"[6]

Today, E10 is a widely accepted motor fuel. During the 1970's, however, questions arose about ethanol's effect on engines, especially relating to phase separation. Fuel tanks would sometimes have a substantial amount of water at the bottom, collected over time because gasoline cannot hold much water (0.15 teaspoons per gallon of fuel at 60°F). In ethanol-free gasoline, even a small amount of water separates and falls to the bottom. Ethanol can absorb and hold much more water (four teaspoons per gallon of E10 fuel). But when an ethanol blend is put in a tank where a large amount of water has collected (more than the ethanol can absorb), some of the ethanol separates from the gasoline along with the water. Today's fuel handling procedures make phase separation quite rare.[7]

Ethanol's ability to hold water is an advantage today. Continued use of E10 can help prevent buildup of water in the fuel tank. Ethanol effectively absorbs small amounts of water and carries it out as the fuel is consumed. Older cars not previously fueled with ethanol blends may have large amounts of water in the tank, in which case a mechanic can drain the tank and refill with an ethanol blend. Phase separation becomes a problem only when a large amount of water is present. The ability to absorb water also enables ethanol to act as a gas line antifreeze.[8]

Plugged fuel filters were another early concern. Ethanol improves engine performance by cleaning the fuel system, but this means a large amount of residue may be dislodged when ethanol is first used in older cars. Changing fuel filters usually solves the problem and continued use of an ethanol blend will prevent another buildup of residue.[9]

Automakers have been designing cars to run on E10 since at least the 1980's. Antique cars may contain parts incompatible with ethanol. The ARCO gas company advises consultation with the manufacturers for information about ethanol compatibility with antique auto parts.[10]

E10 Price and Fuel Economy

Due to tax breaks and sometimes lower production costs, ethanol is usually less expensive than gasoline. Not only is ethanol less expensive, but it also adds 2.5–3 octane points (a measure of engine knock resistance[11]) to E10 gasoline.[12] Ethanol can be an ingredient in low octane gasoline by using a low octane blendstock (base gasoline). More often, though, ethanol is used in higher-octane gasolines (mid-grade or premium).

During 2005, 2–4 cents was the normal discount for E10 according to Ron Lamberty of the American Coalition for Ethanol.[13] In Missouri during 2007, E10 could be had for as much as 10 cents less per gallon than the least expensive ethanol-free gasoline.

Since E10 causes little or no loss in fuel economy in many vehicles, a lower price can mean actual savings. At the 10% level, the favorable combustion properties of ethanol more or less compensate for its lower energy density. A 1998 Study by the U.S. EPA showed an average 1.64% *gain* in miles per gallon for E10 as compared to ethanol-free gasoline. Tests were performed at 0°F and 75°F on 11 vehicles, model years 1977 to 1994.[14] The American Coalition for Ethanol conducted a fuel economy study of three passenger cars, model year 2005. They detected an average 1.47% lower fuel economy for E10 as compared to unleaded gasoline without ethanol.[15] This translates to an average 0.41 fewer miles per gallon—less reduction than you would expect based on energy density alone.

E10 Availability

At least four U.S. states and one Canadian province mandate ethanol in gasoline. Other states are considering mandates. Since 1997, Minnesota has required all gasoline sold in the state to contain 10% ethanol (E10).[16] Hawaii requires 85% of the state's gasoline to contain 10% ethanol. An exemption is allowed if competitively priced ethanol is unavailable or in the case of "undue hardship."[17] A Missouri law requires ethanol in all but the premium grade of gasoline starting January 1, 2008. A special provision lifts this requirement whenever ethanol is more expensive than fossil fuel-based gasoline.[18] In 2005, the governor of Montana signed a law requiring nearly all gasoline sold in the state to contain 10% ethanol beginning one year after in-state ethanol producers achieve the ability to produce 40 million gallons per year and maintain that level for three months.[19] Ontario, Canada's most populous province, mandates ethanol in gasoline as of January 1, 2007.[20] Even in states without mandates, E10 availability is growing.

As ethanol use becomes routine, many states no longer require labels on pumps dispensing E10. If a high-octane gasoline is priced equal to or less than lower grades, it probably contains ethanol. Ethanol tends to bring down the price and adds octane to the base gasoline at the same time!

Running on E85

Slightly modified cars, trucks, and vans known as "flex-fuel vehicles" can burn up to 85% ethanol. Fuel economy goes down, but the E85 optimization technology described in Chapter 6 could reduce that drawback, opening the door to significant savings for motorists. E85 is made of 85% denatured alcohol and 15% gasoline. Flex-fuel vehicles run on E85, E10, ethanol-free gasoline, or any combination of the three. In other words, flex-fuel vehicles can have anywhere from 0% to 85% ethanol in the tank and run perfectly!

The high percentage of ethanol in E85 maximizes benefits from ethanol's low toxicity. The EPA notes E85 is "safer than gasoline to store, transport, and refuel. Because ethanol is water soluble and biodegradable, land and water spills are usually harmless, dispersing and decomposing quickly; the gasoline portion of a spill is still a problem in these situations."[21]

Using E85 also improves air quality. Evaporative emissions straight from the fuel tank, a significant pollution source in warm weather, are substantially reduced with E85 because ethanol has fewer highly volatile components.[22] The U.S. Department of Energy offers these details on E85 exhaust emissions:

> Compared with gasoline-fueled vehicles, most ethanol-fueled vehicles produce lower carbon monoxide and carbon dioxide emissions and the same or lower levels of hydrocarbon and non-methane hydrocarbon emissions. Oxides of nitrogen (NOx) emissions are about the same for ethanol and gasoline vehicles. E85 has fewer highly volatile components than gasoline and so has fewer evaporative emissions.[23]

Flex-Fuel Vehicles

Flex-fuel vehicles (FFVs) are a breakthrough for the "chicken and egg" dilemma hampering renewable fuels. Nobody wants a vehicle they can't use for lack of fuel. Retailers, on the other hand, are not likely to invest in tanks and pumps for a new fuel until a substantial number of motorists can use it. Flex-fuel vehicles are not dependent on one type of fuel and don't usually cost more, allowing their numbers to grow even where E85 is less available. With millions of flex-fuel vehicles on the road and more to come, fuel retailers are steadily adding E85 fuel pumps. Biodiesel, usually made from plant-derived oils, is another renewable liquid fuel that addresses the "chicken and egg" problem.

Over 5 million E85-capable flex-fuel vehicles were on U.S. roads by 2005.[24] Just about every type of automobile is available with the flex-fuel option, including many popular models. These vehicles run perfectly with anywhere from 0% to 85% ethanol in the tank. Flex-fuel vehicles were developed to make use of E85, but run fine on standard gasoline if E85 is unavailable. The technology is becoming routine. The U.S. Department of Energy explains, "FFVs have been used in private and government fleets for years. The technology is proven, and the knowledge base about them is strong. Manufacturers stand behind them with standard warranties equal to those of gasoline vehicles. Dealer maintenance practices for FFVs are very similar to those followed for gasoline vehicles."[25]

Flex-fuel models vary in the way they adjust to different fuels. In general, a sensor automatically detects the ethanol percentage in the fuel and sends a signal to adjust fuel injection and spark timing accordingly.[26] This is necessary mainly because ethanol has a higher oxygen content and lower energy density compared to standard gasoline.

E85 can have a corrosive effect on materials commonly used in fuel systems. FFV gas tanks, fuel lines, fuel injectors, and other parts in contact with fuel must be made of ethanol-compatible materials.[27] Auto mechanics must use ethanol-compatible replacement parts for FFV fuel systems.[28] Some owner's manuals also call for different engine oil.[29] Ford Motor Company stipulated special motor oil when flex-fuel vehicles first became available, but they no longer do so.[30]

Cold weather starting is an issue when using E85 because of ethanol's lower volatility. This is one of the reasons the E85 standard was developed. The 15% non-ethanol portion of E85 helps the engine start in cold weather without pre-heaters or other extra equipment. The ethanol content of E85 is lowered during the winter in cold climates to further help with starting. The 2006 Ford Taurus owner's manual advises the use of non-E85 gasoline or an engine block heater in extremely cold weather if starting becomes a problem. In fact, the Taurus manual recommends the use of an engine block heater regardless of fuel type when temperatures dip to -10°F.[31]

Purchasing a Flex-Fuel Vehicle

The National Ethanol Vehicle Coalition (NEVC) maintains an updated list of flex-fuel vehicles at:

www.e85fuel.com

Flex and non-flex vehicles look the same on the outside. A 2005 survey by South Dakota-based VeraSun Energy found that 68% of flex-fuel vehicle owners do not know they own a flex-fuel vehicle.[32] Some models made since 1998 are flex-fuel as a standard feature for every vehicle of that model rather than just as an option.[33]

How do you know if your particular car has the flex-fuel option? First, locate the Vehicle Identification Number (VIN). The eighth character in the VIN reveals whether or not a vehicle is flex-fuel. Visit the NEVC web site at

www.e85fuel.com/information/vin.php

for a list of characters indicating the flex-fuel feature for each different model.

On newer GM vehicles, check the gas cap. Starting with the 2006 model year, General Motors began putting yellow gas caps on FFVs.[34] Some flex-fuel vehicles are also marked as such inside the fuel filler door. You could also try the owner's manual or check with your dealer.

Having been available in significant numbers for several years, flex-fuel vehicles are showing up on the used car market. Some government and business fleets took advantage of the flex-fuel option early on. Various federal, state, and local government surplus outlets and auctions are sources of used flex-fuel vehicles. We found a 1997 flex-fuel Ford Taurus, for instance, at the Kansas state surplus outlet in Topeka. The U.S. Government Services Administration leases vehicles to federal government agencies. Their web site lists various auctions featuring used FFVs:

www.autoauctions.gsa.gov

Since flex-fuel vehicles accept standard gasoline and are available at little or no extra cost, there is little downside to choosing the option even if you can't always find E85 fuel.

E85 Availability

Stations selling E85 fuel are scarce in many places, but numbers are growing. Go to www.e85fuel.com for a searchable database of E85 stations provided by the National Ethanol Vehicle Coalition. The U.S. DOE also maintains a web site for locating many different alternative fuels at:

www.eere.energy.gov/afdc/infrastructure/locator.html

E85 fuel pumps are scattered across the United States and Canada, with the biggest concentration in Corn Belt states. Minnesota has more E85 pumps than any other state. E85 pumps and ethanol biorefineries are beginning to show up outside the Corn Belt as well.

E85 requires a dedicated storage tank at fuel stations. This can be expensive, but tax incentives and grants are available. Some stations have replaced underused premium gasoline with E85, avoiding the cost of a new fuel storage tank.[35] A European system allows storage of 100% denatured ethanol in one tank and ethanol-free gasoline in another. The two are blended in the desired proportion as fuel is pumped into a vehicle. With this kind of system, only two tanks would be need for dispensing three products—E10, E85, and Ethanol-free gasoline. Such a system would also be ideal for dispensing E20 or E100.[36]

E85 Cost and Fuel Economy

There are many environmental, health, security, and economic reasons for Americans to use E85, but saving money at the pump is usually not yet among them. Cost per mile driven is currently a barrier to E85 expansion. This barrier could come down if new cars are designed for better fuel economy on E85 (see chapter 6) and production costs come down (see chapter 8).

E85 usually costs less than gasoline per gallon, but gets fewer miles per gallon. Lower fuel economy can negate the lower cost per gallon, resulting in a higher cost per mile driven. Figure 5-1 gives average U.S. gasoline and E85 prices as well as the energy equivalent price for E85. The energy equivalent price takes into account the lower BTU content of E85. It reflects the actual cost of driving today's U.S. FFVs. The energy equivalent price could eventually become irrelevant if FFVs are optimized for E85. This is because FFVs optimized for E85 could go more miles on less energy content—more miles per BTU. But for now, figure 5-1 makes it clear that, on average, it is more expensive to use E85 in U.S. FFVs as compared to using ethanol-free gasoline.

How do you know if E85 is a better buy than gasoline for your particular FFV? The percentage price reduction for E85 needs to be greater than the percentage reduction in fuel economy when using E85 vs. standard gasoline. The U.S. EPA and DOE provide an automated comparison tool. Click the cost calculator link at:

www.fueleconomy.gov/feg/flextech.shtml

Select a vehicle and a state. The calculations are done for you. Based on fuel economy reduction for your vehicle and the current average prices for E85 and standard gasoline in your multi-state region, the site tells you what your driving costs will be. In addition, you can find out how many gallons of (ethanol-free) gasoline you will save by using E85 and how much reduction in greenhouse gas emissions will result.

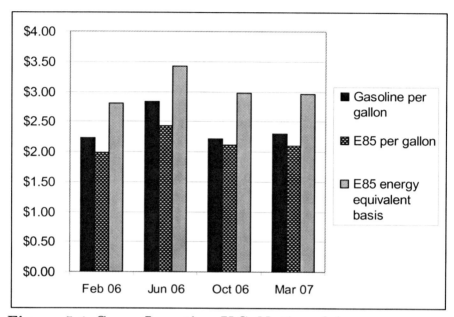

Figure 5-1: Same Location U.S. National Average Retail Gasoline and E85 Prices, 2006–2007 (Data from U.S. DOE Clean Cities Alternative Fuel Price Reports[37])

For a comparison of driving costs reflecting local E85 and standard gasoline prices rather than region-wide averages, follow these steps:

1. Determine your fuel economy (miles per gallon) on E85 and standard gasoline by keeping track yourself or look up EPA fuel economy ratings at **www.fueleconomy.gov**.

2. Divide your miles per gallon on E85 by your miles per gallon on standard gasoline to get your percentage fuel economy reduction when using E85. According to the U.S. EPA, the average fuel economy reduction for 2006 FFV models running on E85 is about 25%.

FIVE | E10, E85, AND FLEX-FUEL VEHICLES   61

3. Check the price of standard (non-E85) gasoline in your area.

4. Take a look at figure 5-2. In the left column, find the number closest to your non-E85 gasoline price. In the top row, find the number closest to your percent reduction in fuel economy when using E85. Where the row and column meet, you will find the approximate E85 price necessary in order to break even on your cost per mile driven. If the price of E85 is less than this break-even point, you will likely save money by using E85 in your FFV. If the E85 price is more than the break-even number, using E85 will probably cost you more. Find the nearest station with an E85 pump by visiting www.e85fuel.com
or www.eere.energy.gov/afdc/infrastructure/locator.html.

Fuel economy is difficult to measure consistently in real-world conditions. Weather, driving style, tire pressure, and engine maintenance all have an effect. That being said, the American Lung Association of Minnesota reports a typical 15%–17% drop in miles per gallon when running on E85 as compared to gasoline in employee driven vehicles.[38] The EPA publishes a comprehensive E85 fuel economy guide for U.S. flex-fuel vehicles. This guide is useful for comparison purposes. Based on EPA figures for all 2006 FFV models, the city/highway average fuel economy for E85 is 25.56% lower than for gasoline. This is because the energy content of E85 is about 27% lower than for ethanol-free gasoline.[39]

	% Fuel Economy Reduction using Alternatives					
	5.00%	10.00%	15.00%	20.00%	25.00% (E85 Avg.)	30.00%
$2.10	$2.00	$1.89	$1.79	$1.68	$1.58	$1.47
$2.20	$2.09	$1.98	$1.87	$1.76	$1.65	$1.54
$2.30	$2.19	$1.07	$1.96	$1.84	$1.73	$1.61
$2.40	$2.28	$2.16	$2.04	$1.92	$1.80	$1.68
$2.50	$2.38	$2.25	$2.13	$2.00	$1.88	$1.75
$2.60	$2.47	$2.34	$2.21	$2.08	$1.95	$1.82
$2.70	$2.57	$2.43	$2.30	$2.16	$2.03	$1.89
$2.80	$2.66	$2.52	$2.38	$2.24	$2.10	$1.96
$2.90	$2.76	$2.61	$2.47	$2.32	$2.18	$2.03
$3.00	$2.85	$2.70	$2.55	$2.40	$2.25	$2.10
$3.10	$2.95	$2.79	$2.64	$2.48	$2.33	$2.17
$3.20	$2.04	$2.88	$2.72	$2.56	$2.40	$2.24

(Standard Gasoline Price)

Figure 5-2: Break-even Prices for Gasoline Alternatives Based on Fuel Economy Reduction

In the near future, innovations in ethanol production and flex-fuel vehicle design could send E85 prices down and fuel economy up. With lower cost and better fuel economy, ethanol will reach its full potential as a renewable energy source beneficial for the economy, environment, national security, and your pocketbook. Following chapters will show how this might happen. With E85 fuel economy equal to or better than that of gasoline, any discount for E85 will be pure savings. A lower E85 price will always be a better deal, and we won't have to get out the calculator to know that's the case!

Notes

1. U.S. DOE National Renewable Energy Laboratory & National Ethanol Vehicle Coalition, *Handbook for Handling, Storing, and Dispensing E85*, 2002, 2.
2. Keith Reid, "Alcohol at the Pump," *National Petroleum News*, (August 2005): 40.
3. U.S. DOE National Renewable Energy Laboratory, *Low-Level Ethanol Fuel Blends: Clean Cities Fact Sheet*, 2005.
4. American Coalition for Ethanol, *Why is Ethanol Good for Your Car?*, http://www.ethanol.org/brochures.html.
5. U.S. Environmental Protection Agency Transportation and Regional Programs Division, *Clean Alternative Fuels: Ethanol*, 2002.
6. U.S. DOE National Renewable Energy Laboratory, *Low-Level Ethanol Fuel*.
7. Facts on phase separation from Downstream Alternatives, *Changes in Gasoline III*, 17.
8. Renewable Fuels Association. *Ethanol and Engine Performance*, 2000.
9. Clean Fuels Development Coalition, *Ethanol Fact Book*, 2003, 42–43.
10. BP West Coast Products LLC, *Frequently Asked Questions*, 2005, http://www.arcogas.com/gas/faqs.php.
11. John D. Heywood, *Internal Combustion Engine Fundamentals* (New York: McGraw-Hill, 1988), 852.
12. Reynolds, *Gasoline Ethanol Blends*, 3.
13. Nate Jenkins, "Buying E10 May Not Always Save Money," *Lincoln Star Journal*, October 14, 2005.
14. Based on tests reported by KT Knapp, FD Stump, & SB Tejada, "The Effect of Fuel on the Emissions of Vehicles over a Wide Range of Temperatures," *Journal of the Air & Waste Management Association* 43 (July 1998).
15. Based on tests by the American Coalition for Ethanol, *Fuel Economy Study*, 2005, http://www.ethanol.org/documents/ACEFuelEconomyStudy.pdf.
16. Office of the Governor of Minnesota, "Governor Pawlenty Signs Nation-Leading Initiative to Double Ethanol in Gasoline by 2013," news release, May 10, 2005.
17. U.S. Department of Energy, *Hawaii Governor Implements Ethanol Requirement*, September 22, 2004, http://www.eere.energy.gov/states/state_news_detail.cfm/news_id=9186/state=HI.
18. Renewable Energy Access, *Missouri Passes Renewable Fuels Act*, July 18, 2006, http://www.renewableenergyaccess.com/rea/news/story?id=45460.
19. U.S. DOE Alternative Fuels Data Center, *Montana Mandates E10 Throughout State*, 2005.
20. Ontario Ministry of Agriculture, Food and Rural Affairs, *McGuinty Government Takes Next Step on Cleaner Air*, October 2005.

21. U.S. Environmental Protection Agency, *Clean Alternative Fuels: Ethanol,* 2002.
22. U.S. Department of Energy, *Handbook for Handling,* 7–8.
23. U.S. Department of Energy Alternative Fuels Data Center, *What Types of Vehicles Use Ethanol?*, 2005, http://www.eere.energy.gov/afdc/afv/eth_vehicles.html.
24. Tara Baukus Mello, "Fueling up With Ethanol," *Edmunds.com,* February 7, 2006, http://www.edmunds.com/.
25. U.S. Department of Agriculture National Renewable Energy Laboratory, *Do You Own a Flex-Fuel Vehicle? Alternative Fuel Fact Sheet,* 2003.
26. Ibid.
27. National Ethanol Vehicle Coalition, *2005 Purchasing Guide for Flexible Fuel Vehicles,* 2005.
28. U.S. Environmental Protection Agency, *Clean Alternative Fuels: Ethanol.*
29. U.S. Department of Energy, *What Types of Vehicles Use Ethanol?.*
30. U.S. DOE Energy Efficiency and Renewable Energy. "Ohio's First Ethanol-Fueled Light-Duty Fleet," *Clean Cities Alternative Fuel Information Series,* May 1999.
31. Ford Motor Company, *2006 Ford Taurus Owner's Manual* (2005), 143.
32. Reid, "Alcohol at the Pump," 41.
33. National Ethanol Vehicle Coalition, *2005 Purchasing Guide.*
34. National Ethanol Vehicle Coalition, "General Motors Leads the Way with E85," news release, September 2005.
35. Reid, "Alcohol at the Pump," 42.
36. Ibid., 41–42.
37. U.S. DOE Clean Cities Alternative Fuel Price Reports, http://www.eere.energy.gov/afdc/resources/pricereport/price_report.html.
38. Don Davis. "E85: Usually cheaper and safer for the environment than gas, but mileage and power may suffer," *The Fargo Forum,* October 23, 2005.
39. National Ethanol Vehicle Coalition, *E85 and Energy Content— how much energy is there?*, http://www.e85fuel.com/.

Chapter 6

IMPROVING FUEL ECONOMY ON ETHANOL

One criticism of ethanol is that it replaces only a small portion of our gasoline use. But this is a poor reason to abandon ethanol. It would make no more sense than abandoning wind power because it can't supply all our electricity. It is, however, a reason to work on conservation and energy efficiency just as intently as production. We use more energy than necessary, primarily because petroleum has historically been inexpensive. As cheap oil becomes scarce, we must find ways to curtail our use of transportation fuels.

Some ethanol critics cite its poor fuel economy compared to gasoline. Actually, ethanol can be an ally in energy conservation and better fuel economy. Preliminary studies show E10 already achieves better fuel economy than ethanol-free gasoline in some vehicles. E85 exhibits reduced fuel economy in today's flex fuel vehicles, but critics often fail to acknowledge this situation can improve. Manufacturers could optimize vehicles for better fuel economy on E10 and E85. This would reduce our cost per mile driven. It would also improve ethanol's energy balance, in turn benefiting our environment, energy security, and trade balance.

Ethanol's energy density is low compared to gasoline, but energy density is not the only property affecting fuel economy. We must also consider how efficiently a vehicle uses energy. No technology is capable of converting 100% of available energy into useful work. Energy is lost to friction and heat. But vehicles optimized for ethanol are able to direct a greater percentage of available energy toward turning the wheels, offsetting lower energy content. This is possible thanks to ethanol's high octane rating and other beneficial combustion properties. Ethanol optimization technologies already exist. Full implementation of these technologies could yield huge benefits through better fuel economy.

In these times of high fuel prices, we often hear of potential solutions such as hybrid vehicles, biofuels, hydrogen, and fuel cells. We need to resist the temptation to see these as competing alternatives. It is tantalizing to think some single "silver bullet" technology will solve all our energy problems. Realistically speaking, if we reject every energy alternative that can't single-handedly replace oil, we will have nothing left. We need many different alternative fuels and most importantly, we need to combine fuels and technologies for maximum efficiency. Biofuels such as ethanol can be a valuable component of a highly efficient transportation system.

Fuel Economy on E10

E10 (10% denatured ethanol and 80% gasoline) works in nearly all gasoline-burning vehicles. In Chapter 5, we introduced two studies indicating little or no reduction in fuel economy on average for E10 as compared to ethanol-free gasoline.[1] The lower BTU content of E10 does not necessarily lead to worse fuel economy. Some cars achieved significantly better fuel economy on E10. Considering most North American ethanol use is currently in the form of E10, its favorable fuel economy redounds to the benefit of ethanol's overall energy balance and cost effectiveness in our transportation system.

Can fuel economy on E10 be improved still more? In both the EPA and American Coalition for Ethanol studies, fuel economy varied considerably among different car models. Some variability would be expected with real world conditions and among different drivers. These studies, however, were done under controlled conditions, indicating some real differences in the efficiency with which different vehicles burn E10.

SIX | IMPROVING FUEL ECONOMY ON ETHANOL

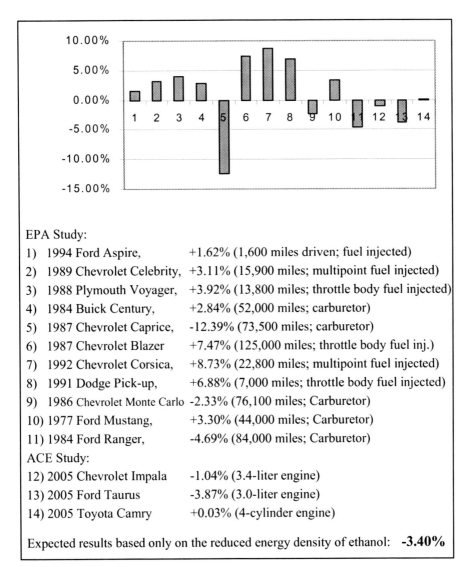

EPA Study:
1) 1994 Ford Aspire,   +1.62% (1,600 miles driven; fuel injected)
2) 1989 Chevrolet Celebrity,   +3.11% (15,900 miles; multipoint fuel injected)
3) 1988 Plymouth Voyager,   +3.92% (13,800 miles; throttle body fuel injected)
4) 1984 Buick Century,   +2.84% (52,000 miles; carburetor)
5) 1987 Chevrolet Caprice,   -12.39% (73,500 miles; carburetor)
6) 1987 Chevrolet Blazer   +7.47% (125,000 miles; throttle body fuel inj.)
7) 1992 Chevrolet Corsica,   +8.73% (22,800 miles; multipoint fuel injected)
8) 1991 Dodge Pick-up,   +6.88% (7,000 miles; throttle body fuel injected)
9) 1986 Chevrolet Monte Carlo   -2.33% (76,100 miles; Carburetor)
10) 1977 Ford Mustang,   +3.30% (44,000 miles; Carburetor)
11) 1984 Ford Ranger,   -4.69% (84,000 miles; Carburetor)
ACE Study:
12) 2005 Chevrolet Impala   -1.04% (3.4-liter engine)
13) 2005 Ford Taurus   -3.87% (3.0-liter engine)
14) 2005 Toyota Camry   +0.03% (4-cylinder engine)

Expected results based only on the reduced energy density of ethanol: **-3.40%**

Figure 6-1: Change in Fuel Economy Using E10 Relative to Ethanol-free Gasoline (data from studies by the EPA and the American Coalition for Ethanol)[2]

Figure 6-1 lists the fuel economy difference for each car as compared to ethanol-free gasoline. Bars extending above the 0% line indicate better fuel economy on E10 as compared to ethanol-free gasoline. Bars below the 0% line indicate worse fuel economy on E10. Of the 14 cars studied, 9 exhibited better fuel economy (more miles per gallon) on E10 as compared to ethanol-free gasoline. More studies are needed on more models, but it appears there is considerable room for designing automobiles optimized for better fuel economy on E10. The fact that some cars already on the road get significantly better fuel economy on E10 than on ethanol-free gasoline suggests E10 optimization is possible without much added expense or new technology.

Simply making E10 performance data available for new and used cars would allow motorists to make buying decisions based on E10 fuel economy. For used cars, the data might not be consistent because differing levels of engine deposits or other use factors might make a difference, but some trend lines might appear even there. New car performance on E10 should be more consistent and measurable.

Improving Flex-Fuel Vehicles

Imagine purchasing a flex-fuel vehicle hoping to save money on fuel and finding out it will actually cost more to drive it on E85 than on standard gasoline. This kind of perceived betrayal doesn't make for happy customers and may cause motorists to shun the whole idea of ethanol fuel. They may even reject E10 under the false assumption that it always lowers fuel economy as well.

A 2006 study by market research company Synovate showed 37% of U.S. consumers were open to buying flex-fuel vehicles. When informed about fuel economy reduction on E85, one third of these same consumers lost interest in flex-fuel vehicles.[3]

If ethanol supporters do not deal squarely and openly with the fuel economy issue, motorists will eventually find out anyway, but in a manner that casts a bad light on ethanol and the ethanol industry. The October 2006 issue of *Consumer Reports* is a perfect example.[4] The cover story, entitled *The Ethanol Myth*, reports a 27% drop in fuel economy experienced by *Con-*

sumer Reports personnel when using E85. The very title of the article casts a negative light on ethanol. To be fair, the story points out fuel economy on E85 could approach that of gasoline if engines were optimized for ethanol. This is the message we need to get across to motorists and automakers, along with the favorable fuel economy results on E10.

Based on EPA figures for all 2006 model year flex fuel vehicles, the city/highway average fuel economy on E85 is 25.56% lower than for ethanol-free gasoline. Some might say this is an unavoidable consequence of the 27% lower energy content for E85 as compared to gasoline, measured in British Thermal Units (BTU's).[5] Many years ago, perhaps this argument was reasonable. Today, however, technologies exist whereby vehicles can be optimized for E85 at a reasonable cost, narrowing the fuel economy gap. The Saab BioPower engines on the roads of Sweden and Great Britain prove this point with their superior power and torque running on E85. Automakers need incentives such as consumer demand or legislation to encourage ethanol optimization of cars sold in North America.

The U.S. Department of Energy is doing their part. In January 2007, they announced availability of grant money for the development of light duty vehicles optimized for ethanol:

> Ethanol has unique properties that are not fully exploited in current technology FFVs. It is the intent of this announcement to undertake research and development projects that will result in FFVs that offer the ability to exploit the favorable properties of the ethanol gasoline blends.[6]

The DOE announced grant recipients later in 2007. An August press release gave details of the seven ethanol optimization projects selected for possible funding:

> Research will seek to take advantage of favorable properties of ethanol blends without diminishing gasoline fuel efficiency. Projects selected for negotiation of awards include:
> Delphi Automotive Systems LLC in Troy, Michigan, has been selected for negotiation of an award of up to $2.2 million for a pro-

ject to demonstrate a vehicle with an E-85 optimized engine, yielding up to 30 percent fuel efficiency improvement. Wayne State University will partner with Delphi.

Ford Motor Company in Dearborn, Michigan, has been selected for negotiation of an award of up to $3.2 million for a project to explore the use of knock-suppression properties of ethanol with increased compression ratios to allow use of smaller, more fuel efficient engines.

General Motors Corporation in Pontiac, Michigan, has been selected for negotiation of an award of up to $1.9 million for a project to develop a cooled exhaust gas recirculation (EGR) combustion prototype, allowing for smaller engines without loss of engine power; this could result in as much as a 15 percent fuel economy improvement. General Motors will partner with Ricardo Inc. for this effort.

Robert Bosch LLC in Farmington Hills, Michigan, has been selected for negotiation of an award of up to $1.5 million for a project to implement an integrated hardware-software system, yielding gasoline-like fuel economy when operating on E-85. Robert Bosch will partner with Ricardo and University of Michigan for this effort.

Siemens Government Services, Inc. in Reston, Virginia, has been selected for negotiation of an award of up to $3 million for a project to investigate the potential of a turbocharged, direct-injection engine operating on E-85 to improve combustion and fuel economy as well as lower exhaust emissions. Siemens will partner with AVL Engineering and Rousch Engineering for this effort.

TIAX LLC in Cambridge, Massachusetts, has been selected for negotiation of an award of up to $1.2 million for a project to develop a novel, high-efficiency engine system for an FFV that not only operates on any blend of ethanol up to E-85, but is projected to exceed the efficiency of a conventional gasoline engine when operated with the highest blends of ethanol (e.g., E85 or higher). TIAX will partner with Monsanto and John Deere for this effort.

SIX | IMPROVING FUEL ECONOMY ON ETHANOL

Visteon Corporation in Van Buren Township, Michigan, has been selected for negotiation of an award of up to $2.3 million for a project to achieve gasoline-like fuel economy when using E-85 by minimizing thermal, dynamic, volumetric, and other system efficiency losses. Visteon will partner with DOE's Argonne National Laboratory, Mahle Powertrain, and Michigan State University for this effort.[7]

The Saab 9-5 BioPower, a car introduced in 2005, was the best selling environmentally friendly vehicle in Sweden by 2006 and accounted for 80% of Saab 9-5 sales there.[8] It is a flex-fuel vehicle, but employs E85 optimization features beyond the flex-fuel technology available in North American vehicles. A Saab dealer's web site explains that turbocharging and engine management technology result in a 20% boost in brake horse power and 16% gain in torque when running on E85 as opposed to regular gasoline.[9]

Please don't miss the significance of ethanol optimization technology. If optimized vehicles were available in the U.S. Corn Belt, it could mean real savings for motorists since E85 is often available at a substantial discount. Motorists driving optimized FFVs would have a stronger incentive to seek out and purchase E85, boosting ethanol consumption. With most FFVs, E85 discounts are rarely steep enough to overcome reductions in fuel economy.

The fuel-saving potential of ethanol optimization technologies could become the main feature in some car models. The high octane of ethanol allows for a greater turbo boost without the pre-detonation (knock) that would otherwise be experienced. Turbocharging is traditionally associated with high performance vehicles, but it could be used in other models as well. Hans Demant, managing director of General Motors subsidiary Opel, was interviewed on this subject for an article in *Automotive News*. He indicated it would be possible to equip other vehicle models with powertrains like the one used for the Saab 9-5 BioPower.[10]

In car models emphasizing fuel economy, a reduction in engine size could partially offset the cost of adding a turbocharger. Because of the added power and torque afforded by turbocharging, these smaller engines could deliver the same performance as larger non-turbo engines. Power and torque would be reduced when lower-octane gasoline is used, but this drawback will fade away as E85 availability improves. Even better, a small turbo-boosted flex fuel engine optimized for ethanol could be combined with an electric motor. Production of this kind of vehicle is technically feasible today.

Ethanol and Hybrid Electric Vehicles

Today's hybrid vehicles combine batteries, an electric motor(s), and a gasoline engine for increased fuel economy. The electric motor is super-efficient around town, while the gasoline engine facilitates longer trips and faster speeds. The engine on such a vehicle could be flex-fuel—one of the standard flex-fuel engines now available in North America, the Saab BioPower engine, or a super-high-efficiency prototype like the MIT direct injection concept described later in this chapter. E85 would be a third fuel choice in addition to gasoline and electricity. Engines could also be designed to run on other biofuels such as biodiesel or butanol (another type of alcohol).

No ethanol-gasoline-electric vehicle is yet available from major manufacturers, but Ford and Saab (General Motors) have featured the concept in prototype models.[11] Toyota, a leader in gasoline-electric hybrid technology, has expressed interest in integrating hybrid technology with alternatives such as ethanol and biodiesel.[12]

Adding plug-in capability could be another step toward better fuel economy. Batteries on a plug-in ethanol-gasoline-electric car could be charged overnight, extending range on electric power alone. This would take advantage of the excess electric power usually available at night when demand is lower.

General Motors may be closest to commercializing an ethanol-capable plug-in hybrid vehicle.[13] Their Chevrolet Volt concept car was unveiled at the 2007 North American International Auto Show in Detroit. The stylish 4-door sedan will go 40 city miles on battery power alone, at which point the super-efficient 1-Litre, 3 cylinder, turbo-charged internal combustion engine will begin to generate electricity. The result will be an amazing 640 mile range without refueling and the equivalent of 150 miles per gallon of gasoline when driven 60 miles per day. Even if you forget to plug in your car at night, you would still get 50 miles per gallon as the engine converts liquid fuel into electricity. The engine can run at a constant "rotations per minute" (rpm) because it is limited to making electricity rather than directly powering the wheels. An engine running at a constant rpm can be designed to operate with greater efficiency.

The Chevrolet Volt sedan will be the first in a series of models using this new propulsion system dubbed "E-Flex" by General Motors. Various engines could be used, powered by fuels such as hydrogen, biodiesel, E85, or even E100 (100% ethanol). E100 could be hydrated (contain water) which would lower the cost and environmental impact of ethanol production.

Advanced, high-capacity batteries at a reasonable cost are crucial for plug-in capability. In May 2007, GM announced two companies have been contracted to develop and supply lithium-ion batteries for the E-Flex system.[14]

The superior range of a plug-in hybrid vehicle would not be possible with electric propulsion alone. The combination of large batteries and a small engine maximizes range and efficiency. Adding the plug-in feature not only takes advantage of off-peak electrical power, it also taps the huge diversity of methods for generating grid-delivered electricity. Ethanol can play a role there as well. Brazilian biorefineries fed by sugar cane have been burning leftover bagasse to produce excess electricity for years. Future cellulosic biorefineries might be capable of producing excess electricity by burning leftover lignin. By feeding this electrical power to the grid, an ethanol biorefinery could, in effect, provide both plug-in electricity and liquid fuel for a plug-in hybrid vehicle.

Ethanol Boosting with Direct Injection

Researchers at the Massachusetts Institute of Technology (MIT) have developed a flex-fuel, turbo-boosted engine prototype that promises fuel economy equal to hybrid gas/electric vehicles without requiring expensive batteries or electric motors.[15] Alternatively, this concept could conceivably be used in combination with electric motors and batteries for even better efficiency.

By direct injection of a small amount of ethanol from a second fuel tank directly into engine cylinders, a high level of knock resistance is created, equivalent to using 130-octane fuel. This allows a high turbo boost and compression ratio, which means a small engine could produce the same power and performance as a much larger standard engine. An engine management system adjusts the amount of ethanol injected. More would be needed at higher engine loads.

With this setup, a 4-cylinder engine could deliver the performance of a much larger 8-cylinder engine. The result would be fuel economy as good as hybrid electric vehicles, but at less initial cost. Over an entire drive cycle, ethanol use could be less than 10% of total fuel use. This small amount of ethanol could serve to reduce overall fuel consumption by 25% compared to a standard engine, while maintaining performance levels. Fossil fuel consumption could be reduced by approximately 31% in this scenario because the ethanol portion is non-fossil fuel. When ethanol is less expensive, it could be used in amounts greater than 10%. E85, already available at some service stations, could be used in place of pure ethanol for direct injection. Even hydrated (wet) ethanol could be used since it would be injected directly into the cylinder and would not mix with gasoline until the last second. Hydrated ethanol is much less expensive and less energy intensive to produce, so it would improve ethanol's energy balance. Perhaps hydrated ethanol could even be marketed in gallon jugs through service stations until a bulk distribution system is built.

Would motorists accept the idea of filling a second fuel tank? According to Ethanol Boosting Systems LLC, a company set up to market this technology, the tank would only need to be filled every 2–4 months.[16] Even if ethanol is unavailable and the second tank is empty, the engine would run, but at reduced peak torque and horsepower. As with other flex-fuel vehicles, the driver would never be stranded for lack of ethanol. On the other hand, this ethanol-boosted engine could be fueled with up to 100% ethanol or E85, taking advantage of those times when ethanol prices are low.

Air pollution levels could be reduced well below the levels of standard gasoline engines because of the high efficiency of this engine. The extra cost of the second fuel tank, turbocharger, and direct injection system would be partially offset by the small size of the engine. Net additional costs would be about $1000.00. Payback from lower operating costs would take around 3–4 years assuming 2006 fuel prices. This is much quicker than with hybrid electric-gasoline vehicles.

This technology, if fully implemented, could dramatically improve ethanol's energy balance, environmental impact, and value to the consumer. The MIT researchers estimate using this engine could result in 7.5 units of useful energy for each unit expended in ethanol production—much better than with current flex-fuel vehicles. All the components of this system are available, ready to be assembled and mass-produced. In fact, Ethanol Boosting Systems is working with Ford Motor Company to develop and test this concept. According to an October 2006 article published by MIT, ethanol-boosted vehicles could be on the road within 5 years.[17]

A Missouri company, Ecosense Solutions, is developing a system for adding ethanol/water injection to existing engines.[18] The first application is irrigation. They completed successful tests in 2006 showing better than 60% fuel savings in irrigation pumps and a significant reduction in emissions. The Ethanol Pumper Retrofit Kit will pay for itself in one week of use according to Ecosense design engineer Russel Gehrke. Ecosense might develop a similar system for retrofitting E85 vehicles. Gehrke expects such a system would reduce vehicle emissions by 90% and increase miles per gallon by 60%.

Ethanol, Hydrogen, and Fuel Cells

Hydrogen use is limited by production and transportation costs. It must be stored and transported in a highly compressed form—700 times the normal pressure of our atmosphere.[19] But what if we could store and transport hydrogen in liquid form? Each molecule of ethanol is a compact bundle of six hydrogen atoms, two carbon atoms, and one oxygen atom. Lanny Schmidt, a researcher at the University of Minnesota, may have found an efficient and economical way to access those hydrogen atoms. His prototype ethanol reactor will extract hydrogen from ethanol—even hydrated ethanol.[20]

Fuel cell and hydrogen technologies are not yet ready for mass production, but integration with biofuels might just hasten their arrival. Dr. Sandy Thomas of H2Gen Innovations believes the hydrogen industry must transition from natural gas to biofuels for the production of hydrogen. He estimates that a fuel cell transportation system could be 2.4 times more energy efficient than an internal combustion engine.[21] This could help reduce our dependence on fossil fuels, especially if we use hydrated ethanol.

Hydrated Ethanol

Not all ethanol is created equal. The difference is water content. Anhydrous (dehydrated) ethanol is virtually water-free—99.5% alcohol (less than 0.5% water by volume). Hydrated ethanol (also known as hydrous ethanol) contains more than 0.5% water. Ethanol is rated by "proof." The proof number is twice the percentage of alcohol in the solution. Anhydrous ethanol, then, is 200 proof. Hydrated ethanol is less than 200 proof. 80% ethanol with 20% water, for example, would be 160 proof.[22]

Ethanol must be anhydrous to be successfully mixed with standard gasoline at the 10% (E10) level, and that last bit of water is expensive to remove. According to the Process Design Center, producing hydrated ethanol costs $.20 less per gallon and uses 45% less energy compared to anhydrous.[23] Nearly all fuel ethanol sold in North America is anhydrous.

HYDRATED ETHANOL IN BRAZIL: According to the Brazilian Ministry of Agriculture, Livestock and Supply, ethanol production from the 2004/05 Brazilian sugarcane crop consisted of 8.2 billion liters of anhydrous and 7 billion liters of hydrated ethanol.[24] How can they use all this hydrated ethanol while North America cannot? Automobiles capable of operating on 100% hydrated ethanol have long been available in Brazil. More recently, Brazilians can purchase flex-fuel vehicles capable of using 100% hydrated ethanol, an E20–E25 gasoline/anhydrous ethanol blend, or any combination of the two.[25] For North American ethanol enthusiasts, "any combination of the two" is likely to come as a shock. We've been told hydrated ethanol can never be mixed with gasoline for fear of phase separation. As it turns out, hydrated ethanol and anhydrous ethanol/gasoline blends *can* be successfully mixed in the gas tanks of flex-fuel vehicles, at least with the 20–25% ethanol blends found in Brazil. Speaking at a 2007 University of Missouri symposium, Brazilian ethanol industry executive Dr. Fernando Reinach related how he takes visitors on a road trip, first filling up with a gasoline/anhydrous ethanol blend, driving a while, and then putting in hydrated ethanol and driving some more.[26] They do this routinely in Brazil!

In 2006, about 70% of the vehicles sold in Brazil were flex-fuel.[27] Both anhydrous ethanol/gasoline blends and hydrated ethanol are widely available throughout Brazil.

HYDRATED ETHANOL IN NORTH AMERICA? In North America, 85% is the highest ethanol proportion found at fuel stations. Hydrated ethanol is virtually impossible to find. Press reports about Brazilian ethanol often highlight the huge discount for ethanol at fuel stations—larger than typical discounts for E85 in North America. The low retail cost is partially due to advantages of climate and reduced production costs for ethanol from sugar cane. However, it is also because the 100% denatured ethanol available in Brazil is hydrated, and therefore did not need that very costly production step—complete dehydration. North American drivers could reap the cost benefits of hydrated ethanol as well, no matter what feedstock is used.

According to Kevin Kenney of Grassroots Energy LLC, switching to hydrated ethanol production would increase existing biorefinery capacities by 10% with only minor modifications. Switching a flex-fuel vehicle from E85 to hydrated ethanol would require an inexpensive cold starting kit consisting of a small tank for gasoline to be used only when starting the engine. Kenney has developed a denatured, hydrated ethanol specification and hopes to market hydrated ethanol in North America.[28] This is not as simple as it sounds because of the tremendous inertia for anhydrous blends in North America. Just as gasohol required support to get off the ground in the 1980's, it will probably require regulatory changes and legislative incentives to clear the way for hydrated ethanol in North America.

Brazil's hydrated ethanol contains about 5% water.[29] Some of the concepts described in this chapter such as ethanol injection and fuel cells might exploit the full potential of hydrated ethanol, even with water concentrations above 5%. The overall result would be greater efficiency on both the production and use sides of the ethanol equation.

Hydrated ethanol is just the sort of innovation needed in order to expand ethanol's share of the liquid fuel market. In the short run, farmers could make their own hydrated ethanol on a small scale without the need for costly dehydration equipment. Farmers could even use solar ethanol stills, further lowering process energy input.

HYDROUS E15: Adding small amounts of hydrated ethanol to gasoline can lead to phase separation. Water separates from other fuel components causing rust and poor engine performance. Researchers at the Process Design Center (locations in the Netherlands, Germany, and U.S.A.), however, found that substantial phase separation will not occur when hydrated ethanol is mixed with gasoline at the 15% level.[30] They refer to this combination as "Hydrous E15." This breakthrough was presented at a sustainable mobility conference in November, 2006.[31] It remains to be seen whether or not North American fuel vendors, auto makers, or regulatory agencies will pursue hydrous E15 or other uses of hydrated ethanol.

Making it Happen

Our review of ethanol use technologies leaves us with two important insights:

1. E10 use does not always reduce fuel economy.

2. Ethanol can be a synergistic ally of other fuel economy boosting technologies.

We should not pit various fuel economy technologies and biofuels against one another. In some cases, different technologies will serve different sectors of the transportation market. In other cases, various combinations of ethanol, other biofuels, batteries, electric motors, turbochargers, and fuel cells can produce much greater efficiencies than any one technology could on its own.

If we are truly serious about energy independence, we must ask automakers and fuel vendors for these fuel efficiency technologies. Turbo boosting, trybrids, and direct injection should be doable very quickly. Ethanol-powered fuel cells will take more time.

As motorists, we need to do our part by putting up with slight inconveniences. In the case of FFVs, this means seeking out E85 where available. In the case of a plug-in trybrid, we would also have the option of plugging into an electrical outlet overnight. In the case of direct ethanol injection, we would occasionally need to fill a second tank with ethanol for maximum benefit. A key feature of all these technologies is that motorists are not stranded for lack of ethanol. These vehicles run on ethanol-free gasoline if necessary, or battery power in some cases. The payoff could be huge. Putting up with small inconveniences could mean smaller fuel bills, not to mention environmental and security benefits. Let's make it happen!

Notes

1. KT Knapp, FD Stump, & SB Tejada, "The Effect of Fuel on the Emissions of Vehicles over a Wide Range of Temperatures," *Journal of the Air & Waste Management Association* 43 (July 1998); American Coalition for Ethanol, *Fuel Economy Study*, 2005, http://www.ethanol.org/documents/ACEFuelEconomyStudy.pdf.
2. The three 2005 model year cars were tested by the American Coalition for Ethanol, *Fuel Economy Study*, 2005, http://www.ethanol.org/documents/ACEFuelEconomyStudy.pdf; The remaining cars were tested for the EPA by KT Knapp, FD Stump, & SB Tejada, "The Effect of Fuel on the Emissions of Vehicles over a Wide Range of Temperatures." For the EPA study, data listed is an average of results at 0° F. and 75° F. Expected results based only on BTU content are from Downstream Alternatives, *Changes in Gasoline III*, 18. An expected 3.4% loss in fuel economy for E10 is based on an energy density of 76,100 BTU's per gallon of ethanol.
3. Synovate, "Study Shows Strong Consumer Consideration for Flex Fuel Vehicles Including E85 Plug-in Hybrids Peak Interest," news release, August 16, 2006.
4. Consumer Reports, *The Ethanol Myth*, October 2006, 15–19.
5. National Ethanol Vehicle Coalition, *E85 and Energy Content—how much energy is there?*.
6. U.S. DOE National Energy Technology Laboratory Office of Freedom Car and Vehicle Technologies, *Grant Opportunity Announcement*, January 17, 2007, http://www.grants.gov/.
7. U.S. DOE, "DOE to Provide up to $21.5 million for Research to Improve Vehicle Efficiency," news release, August 7, 2007, http://www.energy.gov/news/5298.htm.
8. Adrian Smith Saab, *Saab launches BioPower*, 2006.
9. Ibid.
10. Jason Stein, "Opel, Saab developing hybrids," *Automotive News,* March 20, 2006.
11. Forbes, *Saab to showcase world's first ethanol hybrid at Stockholm car show*, April 23, 2006, http://www.forbes.com/finance/feeds/afx/2006/03/23/afx2616065.html; Justin

Hyde, "Ford to debut 1st ethanol hybrid," *Free Press Washington Bureau*, January 25, 2006.
12. NewsTarget, *Toyota looking to develop ethanol-powered vehicles for U.S. market*, July 19, 2006, http://www.NewEnergyReport.org/019700.html.
13. General Motors, "Chevrolet Volt—GM's concept electric vehicle could eliminate trips to the gas station," news release, January 7, 2007, http://www.prdomain.com/companies/G/GeneralMotors/newsreleases/20071838357.htm.
14. General Motors, "GM Awards Advanced Development Battery Contracts for Chevrolet Volt E-Flex System," news release, May 6, 2007, http://www.gm.com/company/gmability/adv_tech/100_news/2007_releases/eflex-batteries-060607.html
15. D.R. Cohn, L. Bromberg, and J.B. Heywood, "Direct Injection Ethanol Boosted Gasoline Engines: Biofuel Leveraging For Cost Effective Reduction of Oil Dependence and CO_2 Emissions," *Massachusetts Institute of Technology*, April 20, 2005.
16. Ethanol Boosting Systems LLC, www.ethanolboost.com/Use.htm.
17. Nancy Stauffer, "MIT's Pint-sized Car Engine Promises High Efficiency, Low Cost," *Massachusetts Institute of Technology News Office*, October 25, 2006, http://web.mit.edu/newsoffice/2006/engine.html.
18. Terry Anderson, "Energy Efficiency-Ethanol Breakthrough Could Reduce Costs, Protect Environment," *Midwest Producer*, September, 2006, http://www.midwestbullseye.com/articles/2006/09/06/news/top_stories/top10.txt.
19. Data about hydrogen and fuel cells from Sam Jaffe, "Independence Way," *Washington Monthly*, July/August 2004, http://www.washingtonmonthly.com/features/2004/0407.jaffe.
20. University of Minnesota News Service, "New Reactor Puts Hydrogen From Renewable Fuels Within Reach," news release, February 12, 2004.
21. Dale Hildebrandt, "Ethanol could be star player in hydrogen fuel cell technology," *Farm & Ranch Guide*, February 16, 2006, http://www.farmandranchguide.com/articles/2006/02/16/ag_news/regional_news/news12.txt.
22. Trucking Research Institute & Hennepin County Central Mobile Equipment Division, "Hennepin County's Experience with Heavy-Duty Ethanol Vehicles," *National Renewable Energy Laboratory,* January 1998.
23. Process Design Center, *Ethanol in Gasoline –Hydrous E15*, presented at Conference on Sustainable Mobility, November 21, 2006, http://www.energyvalley.nl/uploads/Mr_Keuken.pdf.
24. Sergio Barros, "Brazil Sugar Annual 2006," *USDA Foreign Agricultural Service GAIN Report Number BR6002,* April 10, 2006.
25. Fernando Reinach, "Biofuels in Brazil: Today and in the Future," Presentation at the *4th annual Life Sciences & Society Symposium* at the University of Missouri, Columbia, March 15, 2007.

26. Ibid.
27. Office of Global Energy Dialogue/IEA, *The Energy Situation in Brazil: an Overview*, May, 2006, http://www.iea.org/textbase/papers/2006/brazil.pdf.
28. Kevin Kenney, *"EM-95"(tm) 95% Ethanol Based-Spark-Ignition Engine Fuel*, http://www.grassrootsenergy.net/.
29. Ibid.
30. Axel Gottschalk, Jo Sijben, Stefanie Tzioti, *New Insights in Gasoline Blending: Hydrous E15*, http://www.kraftstoffe-derzukunft.de/download/061026Hydrous15Abstract.pdf.
31. Process Design Center, *Ethanol in Gasoline –Hydrous E15*.

Chapter 7

FOOD, FARMING, AND LAND USE

We count on agriculture for a reliable and affordable food supply. American farmers have succeeded in this mission. In fact, farmers have been plagued with surplus production and low prices in recent decades. In 2005, farmers were being paid only around $2.00 per bushel of corn even though about 14% of the crop was put through ethanol biorefineries (USDA ERS data). At these incredibly low prices, it was impossible for most corn farmers to make a profit without government subsidies. While most other commodities rise in line with inflation, grain prices have remained low.

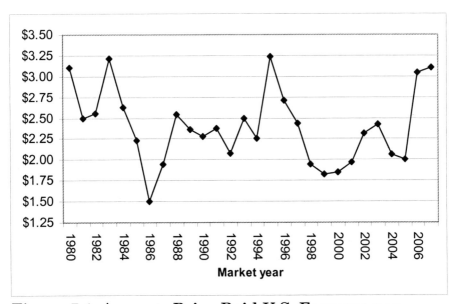

Figure 7-1: Average Price Paid U.S. Farmers per Bushel of Corn, 1980-2007 (data from USDA ERS)

Prices are finally rising, but corn is still a bargain compared to non-food commodities. From the beginning, a major goal of the fuel ethanol movement was higher crop prices so farmers could make a living without government subsidies. That is exactly what is happening.

Growing up in America's Corn Belt during the 1980's, your authors witnessed mountains of corn left in the open many years because large crops overwhelmed grain storage capacity. Grain prices fell, profits evaporated, and many farmers lost land that had been in their families for generations. A few farmers decided it was time to bring back a fuel ethanol industry that could make use of surplus grain. Their hard work eventually paid off. The fuel ethanol industry is thriving while our food supply remains plentiful and relatively inexpensive. But how much can we expect from the land? Will energy production endanger our food supply? Other concerns are even more pressing. The rising price of fossil fuels threatens the profitability of farms reliant on petroleum-derived chemical fertilizers and pesticides. These challenges present opportunities for farmers. Integrated production of food, fuel, and fiber using sustainable farming methods can provide solutions for food and energy security while also improving farm profits and our environment.

Food Prices

Ethanol production did not seem to drive up U.S. food prices significantly through 2006. According to the USDA Economic Research Service (ERS), the consumer price index (CPI) for food increased at an average annual rate of 2.6% from 1996 through 2005.[1] In 2006, with record growth in ethanol production and sharply higher corn prices, the CPI for food went up only 2.3%. This is 0.3% below the average annual increase over the previous 10 years. The CPI for meats alone, which should be more sensitive to the price of corn, went up only 0.7% in 2006.[2] What if corn prices continue to rise? A study by the Center for Agriculture and Resource Development at Iowa

State University estimates a 30% increase in corn prices would boost overall average food prices by only 1.1%.[3]

Corn prices have a minimal effect on the cost of food in part because grain prices are no longer a large component of food prices. The cost of labor is more important today. "Farm value has declined continuously as a percentage of total consumer expenditures, while labor has risen continuously," explains USDA ERS. "Labor is by far the largest and most influential component of the consumer's food dollar. Higher labor costs reflect increased food industry employment stimulated by consumer demand for convenience foods that require significant processing."[4] We also tend to ship foods greater distances than in the past. As the cost of fuel for shipping goes up, so will food prices. Higher crude oil prices tend to boost the cost of goods across the board.

Food AND Fuel from Corn

Running corn through a biorefinery takes only part of it out of the food supply. Ethanol is made from the starch portion of the corn kernel. Starch is less likely than other components to be in short supply for human or animal food. Most of the protein, oil, and other nutrients are left after the starch is removed. These nutrients are available for livestock or human consumption.

About 1/3 of the corn used in dry-mill ethanol production (the most common process) is available in the form of coproduct feeds. This amounts to about 17 pounds of Dried Distillers Grains with Solubles (DDGS) per bushel.[5] DDGS has a greater feed value than the same amount of whole corn in a feed ration. One study showed DDGS has a 27% greater net energy value than dry rolled corn when fed to heifers on forage.[6]

With continued growth in ethanol production, coproduct supplies could eventually exceed the amount needed for animal feed. Coproducts could then be used as fuel for powering ethanol biorefineries. Brazilian biorefineries use leftover portions of sugar cane for power production. Using DDGS this way would reduce the amount of natural gas or coal needed for powering ethanol biorefineries.

Human consumption is another promising use for corn ethanol coproducts. Companies are developing processing techniques that yield corn oil, corn fiber, gluten, and proteins for human consumption, while also producing ethanol and even raw materials for biodegradable plastics. In addition to enhancing our food supply, these new processing methods could increase the profitability of ethanol biorefineries.

EnerGenetics International Inc. of Iowa plans to use small-scale "Mini-Industrial" Biorefineries located close to where crops are grown.[7] They will be able to process specialty products such as organic grains and varieties bred for better yields of particular nutrients. Some of their products are being marketed to the health food industry. Pilot production is underway.

Bio Processing Technology of Indiana will use technology developed at Purdue University.[8] They plan to produce high-value edible products and a specialty protein for making plastics, all while maintaining good ethanol yields. Operating costs will be lower than with the standard wet mill process. Water use, air pollutants, and odor are expected to be much reduced as well.

Ethanol and World Hunger

Understanding the relationship between ethanol and world hunger requires counter-intuitive thinking. Higher grain prices and decreased grain exports do not necessarily exacerbate world hunger. In fact, the opposite could be true in some cases. Most U.S. corn exports go to relatively wealthy countries. Japan topped that list in 2004 (USDA ERS data). Significant amounts also go to less wealthy nations such as Egypt and Mexico. Those with the least income, however, often lack the means to purchase imported grains at any price. Furthermore, corn and other inexpensive commodities imported from the U.S. tend to take away markets for struggling farmers. This can damage rural economies and deepen poverty.

Less than 2% of the national incomes of rich countries come from agriculture. In middle and low-income countries, that figure rises to 17%–35% of gross domestic product.[9] According to the International Food Policy Research Institute, over 75% of the poor in the developing world live in rural areas, and most are farmers. In September 2006, National Public Radio reported U.S. corn exports have damaged the ability of Mexican farmers to make a living. Many of these destitute farmers are attempting to enter the U.S.[10]

In 2007, Alexandra Spieldoch of the Institute for Agriculture and Trade Policy made a statement before the U.S. House Committee on Ways and Means about the consequences of selling grain at low prices in developing countries. "…without substantial governmental support," said Spieldoch, "developing-country farmers are driven out of their local markets by the below-cost imports. …farmers who sell their products to exporters find their market share undermined by the lower-cost competition." According to Spieldoch, agricultural development within developing countries helps drive economic growth. "Research shows," said Spieldoch, "that domestic food productivity is more effective in stabilizing developing-country food security than the reliance on inexpensive (i.e., dumped) food imports. A fair price for the farmer's production will also help stabilize demand for wage labor in the local economy."[11]

Keeping grain prices quite low might seem like a good way to fight poverty, but the opposite result can come about when economies based largely on agriculture are damaged. Ironically, then, a reduction in U.S. exports resulting from increased corn ethanol production might help alleviate poverty-driven hunger in some places when coupled with efforts to enhance food production within developing countries.[12]

When emergency food aide is needed, we might consider sending food products made from Dried Distiller's Grains (DDG's) left over from ethanol production. Ethanol is made from starch, so this is the main corn kernel component missing from DDG's. Starch is probably the least likely component to be lacking in the diets of famine-plagued populations. Indigenous crops such as sweet potatoes and cassava can often supply enough starch in the diet. There is usually more need for fats and proteins, the nutrients found in a concentrated form in DDG's. A given ship or airplane could hold more of these nutrients in the form of DDG's as compared to whole corn kernels. This is because the ethanol-making process has removed much of the starch "filler." DDG's are more perishable than corn kernels, so appropriate processing, storage and transportation techniques would need to be developed for using DDG's in famine relief. DDG's might be incorporated into native foods or easily eaten "food bars" formulated for balanced nutrition. These ideas come from the non-profit Ethanol Producers and Consumers organization. Find them online at:

http://www.ethanolmt.org/

Ready-to-eat DDG-derived foods could actually be better than whole grains in areas lacking the facilities to process and cook whole grain kernels.[13]

Fossil Fuels and Agriculture

Some people oppose corn ethanol because most farmers rely on large amounts of chemical fertilizers, pesticides, herbicides, and diesel fuel for corn production. Many of these fossil fuel-derived inputs contribute to water and air pollution. Row cropping also tends to accelerate soil erosion. These issues must be addressed. Farming efficiency and sustainability can be vastly improved for the benefit of our food supply, environment, and energy production. Those improvements are already being implemented, even in mainstream agriculture.

Cheap oil has made possible profound changes in agriculture. The end of cheap oil will force equally profound changes. In the early 1900's, farms were smaller and more diverse. Small towns across the Midwest were alive with farm-driven commerce. Yields were often lower than today, but were sustained without purchased fertilizers or pesticides.

Concerned about growing world population after World War II, scientists and farm advisors took advantage of new crop varieties, modern chemistry and cheap oil to encourage what came to be known as the "green revolution." Pesticides and fertilizers derived from fossil fuels began to replace animal manures, cover crops, and crop rotations. Petroleum-powered machinery increasingly replaced human and animal power.

Green revolution methods did raise yields, but often at the cost of soil degradation, contaminated water supplies, and petroleum dependence. Over-tillage and excessive applications of nitrogen fertilizers tend to burn up soil organic matter. Organic matter is needed to soak up water and nutrients that are then slowly released for the benefit of soil-building earthworms and microorganisms. This "sponge" effect of organic matter also reduces erosion and dampens the effects of drought. Depletion of organic matter can cause soil erosion, drought stress, and the need for more horsepower to till the ground.

Increasing applications of pesticides and herbicides were required over time, as insects and weeds grew resistant to the chemical onslaught. It was as if our farm fields were addicted to petroleum. The addiction was at its worst during the 1970's. Energy use per unit of agricultural output peaked during the 1970's, declining considerably since 1980.[14] The trend toward greater sustainability in farming continues today, and is one of the reasons corn ethanol requires less fossil fuel input as time passes.

Sustainable Farming

Sustainable farming seeks to preserve the productivity of the land over time while minimizing petroleum inputs. It could also be called high-efficiency or natural farming. Sustainable farming methods are aimed at maintaining the sponge-like organic matter that in turn supports a living ecosystem conducive to crop growth. Some microorganisms fix nitrogen from the air and make it available to plants. Others unlock minerals tied up in clay and rock particles.

Sustainable farming can include no-till and low-till methods that reduce erosion and require fewer trips across fields. New crop varieties deliver higher yields with less fertilization and irrigation. High-tech precision farming uses aerial imagery, extensive soil testing, and satellite-guided application to lower fertilization rates without restricting yields. Fertilizers are placed only where needed and in the amounts needed. Finally, crop scouting helps farmers know when and where to apply pesticides, limiting overall application rates. These tools are widely used and accepted in agriculture today.[15]

A small but growing number of farmers use additional techniques to reduce fossil fuel and chemical use even more drastically while maintaining yields, or even exceeding average yields during drought years. We are not talking about turning the clock back on agriculture. High efficiency farmers combine the best traditional methods, such as crop rotation and cover crops, with state of the art techniques such as ridge till, precision farming, and integrated pest management. Many of these advanced farms are diverse, incorporating grains, grasses, and livestock. Ethanol can fit nicely into such farming systems. Long track records of success for sustainable methods combined with higher energy and chemical costs will continue to drive farmers toward sustainable methods.

Members of Practical Farmers of Iowa (PFI) have used alternative farming techniques for decades with good results. Dick and Sharon Thompson are among the founding members. They helped develop on-farm research methods designed to document and fine-tune techniques such as ridge-till, strip cropping, cover crops, and crop rotation.

In the 1980's and 1990's, conventional cropping systems typically lost money without government subsidies. The Thompsons' have earned good profits from their row crops without subsidies and without the use of herbicides or pesticides for more than 25 years. Compared to conventionally farmed land, their soil has 68 times more earthworms, better structure, and more organic matter. Soil erosion rates are lower as well.[16] Stable soil aggregate, earthworm tunnels, and high organic matter allow soils to soak up more water and resist erosion. Excess moisture is available for later crop use. Thompson corn yields have averaged 155.86 bushels per acre over the years. This is 5.56 bushels better than the county average. During the severe drought year of 1988, per-acre yields were 25 bushels better than the county average.[17]

Whether crops are fed directly to cattle or first processed through an ethanol biorefinery, it's not hard to see how adoption of techniques pioneered by PFI would reduce the need for farm subsidies while improving our environment and farm incomes. Corn ethanol's energy balance would improve as well. According to a 1995 analysis by the Institute for Local Self Reliance, producing enough corn to make a gallon of ethanol would require about 16,800 BTU's worth of energy inputs on the Thompson farm, while the average farm requires 27,134 BTUs. That's a savings of 10,334 BTU's per gallon of ethanol.[18]

A long-term study conducted by the U.S. Department of Agriculture (USDA) confirmed major benefits from one of the techniques used by members of PFI. Cropland managed with ridge tillage for 24 years exhibited less soil erosion and better soil quality as compared to conventional tillage methods.[19]

Using data from published literature, researchers at the University of Michigan found that organic agriculture, which excludes the use of fossil fuel-derived fertilizers, could feed the world without an increase in cultivated area.[20] Leguminous cover crops, they found, can provide enough nitrogen to replace synthetic fertilizers without lowering yields. In fact, they found yields in developing countries could double or triple with a switch to organic methods. In developed countries, yields would be about the same for organic and non-organic farms. The study suggests farm production can be sustained over the long term, even without cheap fossil fuels for fertilization.

Land Use Issues

Through 2006, ethanol production had not trigger a significant trend toward planting more corn acres (figure 7-2). Corn exports had not been curtailed either. In fact, USDA data indicates U.S. corn exports for the 2005 market year reached their highest level since 1995. Surplus production and increasing yield per acre supplied the extra corn needed for ethanol.

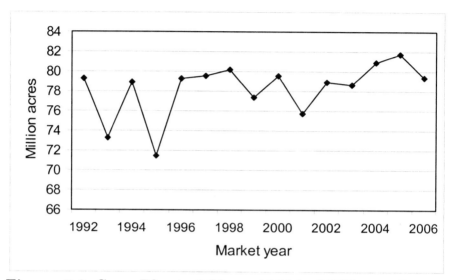

Figure 7-2: Corn Planted Since 1992 in Millions of Acres per Market Year (data from USDA ERS)

If energy prices remain high, ethanol production and corn use will continue to grow. Keith Collins, USDA chief economist, addressed this topic before a U.S. Senate committee in 2006. "Gasoline and ethanol prices," he said, "are likely to stay high enough over the next several years to maintain ethanol expansion."[21]

Cellulosic ethanol from abundant biomass sources could curtail upward pressure on corn prices and acreage, but this will probably not happen for several years. In the meantime, continued yield growth can be expected to supply some additional corn. More corn could also come from a reduction in exports or an expansion in acres planted. According to the USDA, U.S. farmers planted 92.9 million acres of corn in 2007—19% more than in 2006 and the most since 1944.[22]

Diversifying Energy Crops

Even with continuing yield increases, ethanol production could eventually drive up corn prices to the point where many ethanol and livestock producers will suffer. Alternative feedstocks could help moderate corn prices. Legumes, grasses, and other cellulosic feedstocks also improve our soils and environment. Technologies for processing cellulose are just beginning to go commercial (see Chapter 9).

Iowa State University graduate student Andy Heggenstaller is experimenting with alternative crops that could be used for cellulosic ethanol production. One of them is crotalaria, a legume from India. He is also trying rotations of corn with triticale in a single year. Triticale is a cross between rye and wheat. It could produce cellulosic biomass while reducing soil erosion, improving soil structure, and allowing for a corn harvest that same year.[23]

Crop residues such as corn stover will likely be among the cellulosic feedstocks. Crop residues protect soil from erosion and help maintain beneficial organic matter when left on the field. A certain amount can be safely harvested, however, especially if sustainable farming methods are used. Cutting edge farmers such as Richard and Sharon Thompson of Iowa have been harvesting corn stover for animal feed while maintaining soil quality and good yields over a number of years.[24]

A study led by the U.S. DOE National Renewable Energy Laboratory projected Iowa alone could produce close to 2.1 billion gallons of pure corn stover ethanol at prices competitive with the usual corn kernel ethanol. Based on their assumption of no-till farming methods and continuous corn, they found soil erosion would be controlled within tolerable limits established by the USDA, and soil organic matter would remain stable over the 90-year time frame studied.[25] No-till farming is an increasingly common technique using special planting equipment to sow seed through the previous year's stubble.

Soil organic matter, the other variable measured, is an important component of fertile soils. 85% of the crop residue normally left on the ground, says researchers, rots and releases CO_2 into the air. Only 15% is incorporated into the soil as organic matter. That's why 40–50% of the residue can be harvested, at least with no-till farming methods.[26] Essentially, nutrients are being recovered that would otherwise be lost into the air. With no-till farming, less crop residue is needed for the prevention of soil erosion. The USDA's Agricultural Research Service (ARS) is conducting a long-term study to develop more detailed guidelines for sustainable crop residue removal. The study began in 2006 and is scheduled to end in 2011.[27]

Perennials will likely be among the best cellulosic energy crops, living for many years without replanting. Most perennials require little fertilization or tillage. Because the soil is left undisturbed, perennials also serve as ideal hosts for an arbuscular mycorrhizal fungus (AMF) that produces glomalin, a sort of "super-glue" important for soil structure and fertility. AMF grows on plant roots, funneling water and nutrients to the plant while also producing glomalin. ARS researcher Kristine Nichols recently discovered glomalin accounts for a third of the world's stored soil carbon and is an important component of soil organic matter.[28] With extensive root systems, perennial grasses are also exceptionally good at preventing soil erosion.

Perennial and annual legumes such as alfalfa, soybeans, field peas, and various prairie species play host to nitrogen-fixing bacteria in root nodules. These bacteria extract substantial amounts of nitrogen from the air. Generally, legumes require no added nitrogen fertilizer, and even tend to increase soil nitrogen for nearby plants or subsequent plantings. Most nitrogen fertilizers are derived from natural gas. Eliminating the need for nitrogen fertilizers, then, also eliminates a major fossil fuel input.

The ideal energy crop would seem to be perennial legumes mixed with perennial grasses—precisely what the early settlers found growing across vast stretches of North American prairies.

High-Diversity Grassland

In a 2006 research report, David Tilman and others at the University of Minnesota concluded "low-input high-diversity grassland biomass" can be a viable carbon-negative energy crop on highly degraded soils.[29] They found a combination of 16 native grasses, legumes, and forbs grown for combined ethanol and electricity production will yield a net energy balance of 5.1:1 (5.1 units of energy output for every 1 unit of fossil energy input) over a 30 year lifespan. Irrigation was used only for establishment, and no fertilizers were used. The researchers estimate high-diversity grasslands could satisfy 13% of global transportation fuel needs and 19% of electricity needs while occupying only degraded lands no longer suitable for food production. These global estimates are for "integrated gasification and combined cycle technology with Fischer-Tropsch hydrocarbon synthesis." This process results in liquid synfuels and the production of electricity.

Biomass yields increased over time in the Minnesota trials, indicating true sustainability. During the last three years of a ten year sampling period (1996–2005), 16-species plots produced 238% more bioenergy than did monoculture (single-species) plots. Grassland expansion would curtail soil erosion, improve water quality and reduce the amount of topsoil ending up in the Gulf of Mexico. Atmospheric carbon dioxide (a greenhouse gas) could be reduced as well. Carbon is sequestered (trapped) in the soil and roots of high-diversity plots at a much higher rate than for monoculture plots. Total soil nitrogen increased by 24.5% in these low-input, high-diversity plots. This is without fertilization. Total soil nitrogen was unchanged for monoculture plots.

High-diversity, low-input prairie plantings can restore degraded soils while also yielding biofuels. This is incredibly significant. Not only can biofuels be produced without hurting our ability to feed the world, but they can supply the incentive to restore worn-out land that is currently useless for food production. In the process, soil erosion and greenhouse gases will be reduced and wildlife habitat enhanced without the need for irrigation or fossil-based fertilizers and pesticides.

Enhancing Food Production with Energy Farming

If done properly, cellulosic ethanol could actually enhance soil fertility and food production capacity. Soil degradation is a threat to world food production, wildlife, and water quality. Our life-support system depends on a thin layer of fertile topsoil. Through the Conservation Reserve Program and other mechanisms, we have been paying farmers *not* to cultivate highly erodable and otherwise fragile soils. By planting a wide variety of native prairie plants, fertility could be restored to these worn-out soils while also producing energy. After a number of years, some of this land could be converted grazing or other types of food production, creating long-term rotations.

Grasses are beneficial in crop rotations. USDA and Auburn University researchers found that including switchgrass in a rotation with peanuts effectively controls harmful nematodes, eliminating the need for toxic nematicides.[30] Lee Lynd of Florida State University notes a rotation of switchgrass and corn would allow harvesting a greater portion of the stover without depleting soil carbon.[31]

Principles of biodiversity and mimicking native biospheres could be applied to various regions and soils. Acidic soils, for example, are suited for native trees and woody perennial crops. Some portions of the Great Plains will do better with perennial species accustomed to less rainfall, such as prairie cordgrass.

The experiences of native seed producers and prairie enthusiasts will be valuable in selecting the best mix of species and management practices in various regions. Many landowners have established native plantings for the benefit of wildlife. The same biodiversity that appears to be good for bioenergy production is also ideal for quail and other wild animals. There are plenty of reasons for landowners, environmentalists, hunters, economists, and farmers to support cellulosic biofuel production.

Notes

1. USDA Economic Research Service, *Food CPI, Prices, and Expenditures,* March 26, 2007, http://www.ers.usda.gov/.
2. USDA Economic Research Service, *Food CPI, Prices, and Expenditures: CPI for Food Forecasts,* March 26, 2007, http://www.ers.usda.gov/.
3. Helen H. Jensen and Bruce A. Babcock, "Do Biofuels Mean Inexpensive Food Is a Thing of the Past?," *Iowa Ag Review Online* 13 (Summer 2007).
4. USDA Economic Research Service, *Food Marketing and Price Spreads: Relationships between Price Spreads and Marketing Input Costs,* April 25, 2002, http://www.ers.usda.gov/.
5. Keith Collins (Chief Economist, USDA), *Statement before the U.S. Senate Committee on Environment and Public Works,* September 6, 2006, http://www.usda.gov/oce/newsroom/congressional_testimony/Biofuels%20Testimony%209-6-2006.doc.
6. G. E. Erickson, T. J. Klopfenstein, D. C. Adams, and R. J. Rasby, "General Overview of Feeding Corn Milling Co-Products to Beef Cattle," in *Corn Processing Co-products Manual* (University of Nebraska Lincoln), January 2005, http://beef.unl.edu/byprodfeeds/manual_02_02.shtml.
7. EnerGenetics International, Inc., *Executive Overview,* 2006.
8. Purdue Research Foundation, "New Ethanol Process Offers Lower Costs, Environmental Benefits," news release, September 15, 2006.
9. Kevin Watkins and Joachim von Braun, "Time To Stop Dumping On The World's Poor," *2002–2003 IFPRI Annual Report,* http://www.ifpri.org/.
10. Lourdes Garcia-Navarro, "Economy, Politics Collide in a Divided Mexico," *National Public Radio's Weekend Edition,* September 16, 2006, http://www.npr.org/.
11. Alexandra Spieldoch, Statement before the U.S. House of Representative Committee on Ways and Means, 2007, http://waysandmeans.house.gov/hearings.asp?formmode=view&id=5641
12. In an address before a general assembly of the United Nations on September 25, 2007, U.S. President George W. Bush proposed purchasing commodities for food aide from farmers located close to where the food is needed. Coupled with educational efforts for farmers, this kind of approach could be more effective in the long run rather than shipping cheap commodities from across the globe.
13. Ethanol Producers And Consumers, *Distiller Grain Co-products,* http://www.ethanolmt.org/dg.html
14. Randy Schnepf, "CRS Report to Congress, Energy Use in Agriculture: Background and Issues," *Congressional Research Service,* November 19, 2004, 7–8.

15. Hosein Shapouri and Michael Salassi, "The Economic Feasibility of Ethanol Production from Sugar in the United States," *U.S. Department of Agriculture*, July 2006, http://www.usda.gov/oce/EthanolSugarFeasibilityReport3.pdf.
16. Richard, Sharon, Rex, and Lisa Thompson, *Alternatives in Agriculture: Thompson on Farm Research, Summary,* 2004, 3–4, http://www.pfi.iastate.edu/ofr/Thompson_OFR/TOC_Thompson.htm.
17. Ibid., *Section 5*, 1.
18. David Lorenz and David Morris, "How Much Energy Does It Take to Make a Gallon of Ethanol?," *Institute for Local Self Reliance*, August 1995, http://www.carbohydrateeconomy.org/library/admin/uploadedfiles/How_Much_Energy_Does_it_Take_to_Make_a_Gallon_.html.
19. C. A. Cambardella, S. S. Andrews, and D. L. Karlen. "Watershed-scale Assessment of Soil Quality in the Loess Hills of Southwest Iowa," *Soil and Tillage Research* 78 (August 2004): 237–247.
20. University of Michigan News Service, "Organic farming can feed the world, U-M study shows," news release, July 10, 2007.
21. Keith Collins, *Statement before the U.S. Senate*.
22. USDA National Agricultural Statistics Service, "U.S. Farmers Plant Largest Corn Crop in 63 Years," news release, June 29, 2007.
23. Charlotte Eby, "Keeping it clean Ethanol industry improves environmental practices," *Sioux City Journal*, October 5, 2006, http://www.siouxcityjournal.com/special_sections/ethanol/.
24. Richard, Sharon, Rex, and Lisa Thompson, *Alternatives in Agriculture: Thompson on Farm Research*.
25. John Sheehan et al, "Energy and Environmental Aspects of Using Corn Stover for Fuel Ethanol," *Journal of Industrial Ecology* 7 (Summer/Fall 2003): 117–146.
26. U.S. DOE National Renewable Energy Laboratory, *Life-Cycle Analysis of Ethanol from Corn Stover*, March 2002, http://www.nrel.gov/docs/gen/fy02/31792.pdf.
27. USDA Agricultural Research Service, *Impact of Residue Removal for Biofuel Production on Soil—Renewable Energy Assessment Project*, 2007, http://www.ars.usda.gov/research/projects/projects.htm?accn_no=410653.
28. Don Comis, "Glomalin: Hiding Place for a Third of the World's Stored Soil Carbon," *Agricultural Research* (September 2002); also at http://www.ars.usda.gov/is/AR/archive/sep02/soil0902.htm.
29. David Tilman, Jason Hill, and Clarence Lehman, "Carbon-Negative Biofuels from Low-Input High Diversity Grassland Biomass," *Science* 314 (December 8, 2006): 1598–1600.
30. Rodrigo Rodriguez-Kabana, "Warm-Season Forage Grasses as Rotations for Sustaining Profitable Peanut Production," *USDA—Sustainable Agriculture Resource and Education*, 1996, http://www.sare.org/
31. Lee R. Lynd, "The Role of Biomass in Meeting U.S. Energy Needs," presentation at *Growing the Bioeconomy* in Ames, Iowa, August 29, 2005.

Chapter 8

ETHANOL PRODUCTION

We should evaluate the ethanol industry not by yesterday's technology, but by the high-efficiency methods rapidly being implemented. Due to rising costs, producers are realizing the need to lessen dependence on corn kernels as a feedstock and natural gas as a process fuel. This chapter is about the production advances making ethanol make sense—making it less expensive and more environmentally friendly. Advances might even involve producing other biofuels such as butanol. These technologies will steadily improve the sustainability of biofuels.

Large Scale Ethanol Production

Ethanol is generally made like beverage alcohol—sugars are fermented and the resulting "beer" is purified by distillation. Another approach is the syngas (or thermochemical) platform. Cellulosic biomass is gasified, producing a synthesis gas which can be converted into ethanol or other useful substances through chemical catalysis.[1] Pyrolysis of cellulosic biomass also has potential. Pyrolysis oil can be upgraded to ethanol or other biofuels.

Ethanol can be made from just about any organic material. Simple sugars allow the easiest path to ethanol production. Feedstocks such as sugar cane, sweet sorghum, Jerusalem artichoke tops, and sugar beets contain readily fermentable sugars.

North America's ethanol industry is currently based on starch feedstocks. Starch must be hydrolyzed, releasing sugars prior to fermentation. It can be stored for long periods, facilitating year-round biorefining. Field corn (not sweet corn) is the main starch crop for making ethanol in North America.

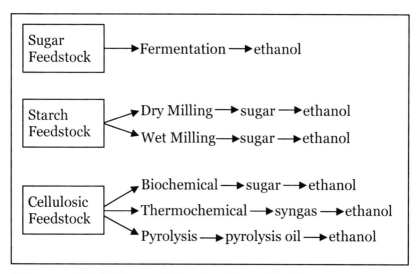

Figure 8-1: Ethanol Production Paths

Cellulosic feedstocks are on the verge of commercialization. Cellulose and hemicellulose are more abundant than starch or free sugars. The use of cellulosic feedstocks could allow a huge expansion in biofuel production and divert waste streams from landfills. However, it is more difficult to liberate sugars from cellulosic materials than from starch. Cellulosic ethanol is being made on a pilot and demonstration scale. Until production costs come down, however, most North American ethanol will be made from the starch found in corn kernels. Even though cellulose will likely surpass corn as an ethanol feedstock some day, we owe a debt of gratitude to pioneers in grain-based ethanol production. As investor Vinod Khosla points out, we would never achieve a cellulosic biofuel industry without the foundation provided by corn ethanol.[2]

EIGHT | ETHANOL PRODUCTION

Feedstock	Capacity (million gallons)	% of Total	Number of Biorefineries
Corn[a]	4,516	92.7%	85
Corn/Milo	162	3.3%	5
Corn/Wheat	90	1.8%	2
Corn/Barley	40	0.8%	1
Milo/Wheat	40	0.8%	1
Waste Beverage[b]	16	0.3%	5
Cheese Whey	8	0.2%	2
Sugars & Starches	2	0.0%	1
Total	4,872	100.0%	102

a/ Includes seed corn
b/ Includes brewery waste

Figure 8-2: Ethanol Production by Feedstock, 2006
(Adapted from U.S. Environmental Protection Agency, Office of Transportation and Air Quality *Renewable Fuel Standard Program—Draft Regulatory Impact Analysis*, September 2006, EPA420-D-06-008)

New Feedstocks

Corn got us started, but we need additional feedstocks adapted to various climates and soils. These might include soybean stalks, field peas, grain sorghum, sweet sorghum, sugar beets, wheat, barley, molasses, cheese whey, Jerusalem artichokes, cattails, tree trimmings, grass clippings, windfall fruit, vegetable and fruit wastes, and all kinds of organic solid waste. The possibilities are seemingly endless. We'll look at some of the promising new feedstocks in addition to field corn and sugar cane—today's leading ethanol crops.

	Feedstock type		Ethanol Yield in
	Sugar	Starch	Gallons/Acre/Year
Sugar Beet[a]	X	X	750
Sugar Cane	X		695 (Florida)[b]
			665 (Texas)[b]
			870 (Brazil)[c]
Sweet Sorghum[d]	X		400–600
Jerusalem Artichoke	X	X	300–600 (from tubers)[e]
			770–941 (from tops)[f]
Field Corn[g]		X	413
Grain Sorghum[g]		X	172
Barley[h]		X	150
Wheat[i]		X	106
Field Pea[j]		X	78
Sweet Potato[k]		X	712–1140

a/ Based on yields in France. Hosein Shapouri & Michael Salassi, *The Economic Feasibility of Ethanol Production from Sugar in the United States*, USDA (July 2006).
b/ Based on 2003–05 yields. Shapouri & Salassi, *The Economic*
c/ For 2006. Milton Maciel, "Ethanol from Brazil and the USA," *ASPO-USA/ Energy Bulletin* (Oct 2006), http://www.energybulletin.net/21064.html.
d/ Based on Iowa yields. I.C. Anderson, *Ethanol from Sweet Sorghum*, Iowa State University.
e/ Based on 14–20 tons of tubers per acre. W.C. Fairbank, *Jerusalem Artichoke for Ethanol Production*, University of California University Extension/Vegetable Research and Information Center.
f/ Based on yields in western Canada. L. Baker, P.J. Thomassin, J.C. Henning, "The Economic Competitiveness of Jerusalem Artichoke (Helianthus tuberosus) as an Agricultural Feedstock for Ethanol Production for Transportation Fuels," *Canadian Journal of Agricultural Economics* 38 (1990), 981–990.
g/ Based on 2003–05 U.S. yields. Shapouri & Salassi, *The Economic*
h/ Based on Southeast U.S. Kevin B. Hicks et. al., *Current and Potential Use of Barley in Fuel Ethanol Production*, 2005 EWW/SSGW Conference.
i/ Based on Canadian yield of 40 bushels/acre. (S&T)2 Consultants Inc., *The Addition of Ethanol From Wheat to GHGenius*, Prepared for Natural Resources Canada Office of Energy Efficiency, (January 31, 2003).
j/ Based on dry seed yield of 46 bushels/acre. Jan Suszkiw, "Imagine—Fuel Alcohol From Pea Starch!," *Agricultural Research Magazine*, USDA ARS (March 28, 2006); Kent McKay, Blaine Schatz, and Gregory Endres, *Field Pea Production*, NDSU Agriculture and University Extension (March 2003).
k/ 1983 estimate for the "HiDry" cultivar. A. Jones, M. Hamilton, P.D. Dukes, "Sweet Potato Cultivars for Ethanol Production," *Proceedings of the Third Annual Solar Biomass Workshop, Atlanta, GA*, (April 26–28, 1983), 195–198.

Figure 8-3: Estimated Ethanol Yield by Feedstock

Ethanol from Sugar Feedstocks

Starch from corn kernels must be broken down for ethanol production. With sugar cane, sweet sorghum, sugar beets, Jerusalem artichoke tops, fodder beets, and some food wastes, this step is unnecessary. Juice laden with simple sugars is pressed from plant materials already in a fermentable state, reducing processing costs. Sugar cane bagasse (leftover material) is usually burned, producing heat and electricity for the biorefinery with some left over for electric utilities.

BRAZILIAN SUGAR CANE: Brazil's successful energy farming system is based on sugar cane. They achieve an average 9:1 energy balance, much better than that of corn kernel-derived ethanol.[3] Ethanol industry executive Dr. Fernando Reinach calculates future ethanol yields could reach over 2,300 gallons/acre/year based on new hybrid sugar cane varieties, cellulose conversion, and improved technology.[4]

U.S. SUGAR CANE: Sugar cane requires a tropical or subtropical climate, found only in far southern areas of North America. Brazil's climate allows for a longer harvest season, keeping ethanol biorefineries busy. They also benefit from a long history of industry development. The USDA believes co-location of ethanol and edible sugar production might be viable in the U.S. They estimate ethanol from molasses, a by-product of the sugar refining process, could be made at a cost of $1.05 per gallon.

> Existing sugarcane mills and sugar beet plants could be modified to produce both sugar and ethanol in the same plants. Rather than build a stand alone plant to convert sugarcane or sugar beets into ethanol, it may be more economical to modify existing plants for the production of ethanol. Less capital investment is required to modify an existing sugarcane or sugar beet mill for production of both sugar and ethanol. The front end of a sugar mill is the same for production of sugar or ethanol. Beet and cane juices are extracted in the first stage of converting sugarcane or sugar beets into either ethanol or raw cane or refined sugar.[5] (USDA)

Cost Item	Estimated Ethanol Production Costs[1]						
	U.S. Corn wet milling	U.S. Corn dry milling	U.S. Sugar Cane	U.S. Sugar Beets	U.S. Molasses [2]	Brazil Sugar Cane [3]	E.U. Sugar Beets [3]
Feedstock costs [4]	$0.40	$0.53	$0.48	$1.58	$0.91	$0.30	$0.97
Processing costs	$0.63	$0.52	$0.92	$0.77	$0.36	$0.51	$1.92
Total cost	$1.03	$1.05	$2.40	$2.35	$1.27	$0.81	$2.89

1/ Excludes capital costs.
2/ Excludes transportation costs.
3/ Average of published estimates.
4/ Feedstock costs for U.S. corn wet and dry milling are net feedstock costs; feedstock costs for U.S. sugarcane and sugar beets are gross feedstock costs.

Figure 8-4: Estimated Production Cost by Feedstock
(U.S. dollars per gallon; adapted from Hosein Shapouri and Michael Salassi, *The Economic Feasibility of Ethanol Production from Sugar in the United States*, USDA, July 2006)

Farms such as Northside Planting LLC of Franklin, Louisiana are developing sustainable farming systems for sugar cane. They have been able to cut fertilizer inputs without sacrificing yield.[6]

Japan's National Agricultural Research Centre for Kyushu Okinawa Region and Asahi Breweries of Japan are developing a giant, high-biomass cane by crossing cultivated and wild strains. It thrives with poorer soils and dry conditions, thanks to a deeper root system.[7] Amazingly, Asahi claims the new variety can produce triple the ethanol while also yielding more sugar and more bagasse for power production as compared to standard strains. It is hoped this new strain will help rescue Japan's sugar industry as it strives to compete with imported sugar. Pilot scale ethanol production from the new variety is scheduled for 2010.[8]

Ethanol from Sweet Sorghum

Sweet sorghum (*Sorghum bicolour*), is valued for large stems with sweet juice, similar to sugar cane. It thrives throughout much of North America and requires less irrigation in dry climates compared to corn. Ethanol yield per acre is better than for corn kernels as well.

Very little ethanol is made from sweet sorghum, largely because of difficulty with transportation and storage. Large biorefineries need a continuous supply of feedstock, but sorghum juice deteriorates rapidly, and is typically harvested only once per year in North America. Solutions for this problem might be close at hand. EnerGenetics International Inc. of Iowa is developing an ensilage system to preserve sorghum cane in huge plastic "bags." The feedstock is then available for processing in mini-biorefineries placed close to the fields, reducing transportation costs. These small biorefineries could be owned and operated by cooperatives or small groups of farmers. They will require no process fuel other than feedstock coproducts. This should bode well for the energy balance of ethanol produced with the system.[9]

Iowa State University has also developed an ensilage system for sweet sorghum. It allows on-farm storage for many months. In a single pass through the field, sorghum cane is harvested, chopped, and sprayed with sulfuric acid and yeast inoculums. The harvest then goes in a covered pit silo. Using long-season southern varieties, Iowa State has achieved sugar yields sufficient for 400–600 gallons of ethanol per acre. In Iowa, long-season varieties have immature heads at harvest, preventing loss of sugar from grain formation. The largest sugar yields were obtained during the hottest, driest growing seasons.[10] Drought usually means lower corn yields, so extra production from sweet sorghum in those years would be welcome.

Another Iowa company is taking a different approach. Lee McClune, president of Sorganol Production Co. Inc., is building a machine to harvest and squeeze juice from sweet sorghum stems as it is pulled through the field.[11] This juice will be fermented and stored in tanks on each farm. A portable distillation unit will be used for conversion to ethanol. This unit could be shared among a number of farmers, moved from farm to farm. The on-farm approach eliminates feedstock transportation costs. McClune is perfecting his system with the help of Oklahoma State University.[12]

Ethanol from Jerusalem Artichokes

Large quantities of easily fermentable sugars can be harvested from the tubers or tops of the Jerusalem artichoke (*Helianthus tuberosus L*), a North American native related to Sunflowers. Abengoa Bioenergy is researching the production of ethanol from Jerusalem artichoke tops/stems rather than roots, avoiding costly digging. They are studying sugar content in the stems to determine the best growth stage for harvest. Sugars are concentrated in the tops before blossoming and in the tubers later on.[13]

A 1990 study estimated 770–941 gallons of ethanol per acre from Jerusalem artichoke tops based on yields in western Canada.[14] These figures rival ethanol yields from sugar cane. Plant breeding and process optimization could improve yields still more.

Another Canadian study, published in 1998, predicted ethanol could be produced from Jerusalem artichoke tops at a cost competitive with corn ethanol. With this plan, farmers throughout Quebec would plant Jerusalem artichokes on subprime land currently used for hay crops. The tops would be delivered to a network of strategically located biorefineries.[15]

If only the tops were used, Jerusalem artichokes could be maintained as a perennial crop, eliminating much of the tillage and soil erosion associated with annual crops like corn. The Jerusalem artichoke is an extremely tenacious, strong-growing perennial, adaptable to many soil types and climates. It thrives on soils too poor for most row crops. Jerusalem artichoke tops look promising for North America, and without the need for cellulosic conversion technologies. In fact, its simple sugars should be easier to convert than corn starch.

Jerusalem artichoke tubers are also high in sugars. According to the University of California Vegetable Research and Information Center, it would be reasonable to expect a yield of 14–20 tons of tubers per acre. With state-of-the-art conversion technology, this would mean 300–600 gallons of ethanol per acre. The UC Davis Experiment Station reported tuber yields up to 33 tons per acre. Dewatered stillage left over from ethanol production is a protein-rich animal feed.[16]

Ethanol from Food Waste

Ethanol production can turn a disposal problem into an asset.[17] Food wastes such as fruit, cheese whey, potato, and beverage waste have enough free sugars or starch to be processed into ethanol using standard low-cost methods. In March 2007, the Renewable Fuels Association listed two U.S. companies producing ethanol from cheese whey and four using beverage waste or waste beer.[18]

SWEET POTATO WASTE: CSEA Co-Operative Inc. of Canada plans to make ethanol from sweet potatoes and sweet potato waste, in addition to other feedstocks such as millet and sorghum. Leftover coproduct will be digested to make a biogas that will be sold as a substitute for natural gas or burned to produce electricity. About 40% of the resulting electrical power will be used at the biorefinery. Excess power, perhaps enough to supply 6000 homes, will be sold through the existing grid.[19]

The farmer-owned CSEA cooperative was formed in a tobacco growing region of Ontario. The declining market for tobacco prompted a search for alternative crops. If the project is successful, it could be worth a try in other tobacco regions.[20]

WATERMELON WASTE: Bob Morrissey, executive director of the National Watermelon Association, advocates ethanol production from unusable watermelons. According to Morrissey, 800 million pounds of watermelon are abandoned in the field each year. Watermelons are about 10–14% sugar, easily convertible to ethanol.[21]

On-Farm Ethanol from Waste Fruit

Daniel West of Macon, Missouri recognized an untapped source of energy in his orchard. The 10 acres of apples, peaches, apricots, nectarines, plums, and pears are a good source of sugar-laden waste fruit. With the help of a grant from the USDA's Sustainable Agriculture Research and Education program, he built a simple on-farm still capable of making 10 ounces of ethanol per minute from waste fruit. The cost comes to about $0.60 per gallon.[22] That's the cost of water and electricity to run the still. He uses the ethanol in tractors and other on-farm equipment.

West's efforts show that economy of scale (getting bigger) is not the only way to reduce cost and energy use for biofuel production. Small-scale production can take advantage of factors such as niche feedstocks, lower transportation costs, and solar distillation. "If any other farmers or orchard owners would approach me and ask the feasibility of this project," says West, "I would encourage them all to try their own project for sustainability on their farm."[23]

Ethanol from Beets

Sugar beets can yield more ethanol per acre than leading ethanol crops (figure 8-2). However, large-scale ethanol production from beets is more costly than from corn or sugar cane, at least in areas best suited for growing corn or sugar cane. Good beet yields are possible in areas too cold for most row crops, however, and beets may be ideal for farm or home production. The high sugar content facilitates a simple conversion process, a definite advantage for small-scale production.

Ethanol from Corn Kernels

In the U.S., ethanol is mostly made from the starch portion of corn kernels. A tiny part is made from other starchy grains using similar methods. The two main processes are dry milling (about 79% of total production) and wet milling (about 21% of total production).[24]

WET MILLING: In wet milling, grains are soaked in water or a dilute acid solution to facilitate separation into different "fractions" such as starch, protein, germ, oil, and fiber. This process is known as "fractionation." The starch portion is broken down into simple sugars by enzymes (enzyme hydrolysis). These sugars are fermented with the help of yeast, producing a "beer" ready for distillation. Non-starch fractions become valuable coproducts—food additives, vitamins, and animal feed.[25]

DRY MILLING: The dry mill process has fewer steps. Corn kernels are finely milled, mixed with water, and heated. The resulting "mash" is subjected to enzyme hydrolysis and fermentation. The U.S. Department of energy describes dry mill ethanol production with the following steps:

Dry Mill Ethanol Production[26]

1. **Milling.** Corn kernels are ground into a fine powder called "meal." Liquefying and Heating the Cornmeal. Liquid is added to the meal to produce a mash, and the temperature is increased to get the starch into a liquid solution and remove bacteria present in the mash.

2. **Enzyme Hydrolysis.** Enzymes are added to break down the long carbohydrate chains making up starch into short chains of glucose (a simple 6-carbon sugar) and eventually to individual glucose molecules.

3. **Yeast Fermentation.** The hydrolyzed mash is transferred to a fermentation tank where microbes (yeast) are added to convert glucose to ethanol and carbon dioxide (CO_2). Large quantities of CO_2 generated during fermentation are col-

lected with a CO_2 scrubber, compressed, and marketed to other industries (e.g., carbonating beverages, making dry ice).

4. **Distillation**. The broth or "beer" produced in the fermentation step is a dilute (10 to 12%) ethanol solution containing solids from the mash and yeast cells. The beer is pumped through many columns in the distillation chamber to remove ethanol from the solids and water. After distillation, the ethanol is about 96% pure. The solids are pumped out of the bottom of the tank and processed into protein-rich co-products used in livestock feed.

5. **Dehydration**. The small amount of water in the distilled ethanol is removed using molecular sieves. A molecular sieve contains a series of small beads that absorb all remaining water. Ethanol molecules are too large to enter the sieve, so the dehydration step produces pure ethanol (200 proof). Prior to shipping the ethanol to gasoline distribution hubs for blending, a small amount of gasoline (~5%) is added to denature the ethanol making it undrinkable.

CORN ETHANOL YIELD POTENTIAL: According to Dr. Martha Schlicher of Renewable Agricultural Energy Inc., there is considerable room for ethanol yield growth from corn kernels. Thanks to improvements in agronomic productivity, ethanol process optimization, and tailored hybrids, Schlicher predicts ethanol yields from corn kernels could reach more than 618 gallons/acre/year by 2015. This would be near a 50% improvement over 2005, when little more than 400 gallons/acre was the norm. According to Dr. Schlicher, 20 billion gallons of ethanol per year could realistically be produced from U.S. corn by 2015, with enough corn kernels left for livestock feed and other uses.[27]

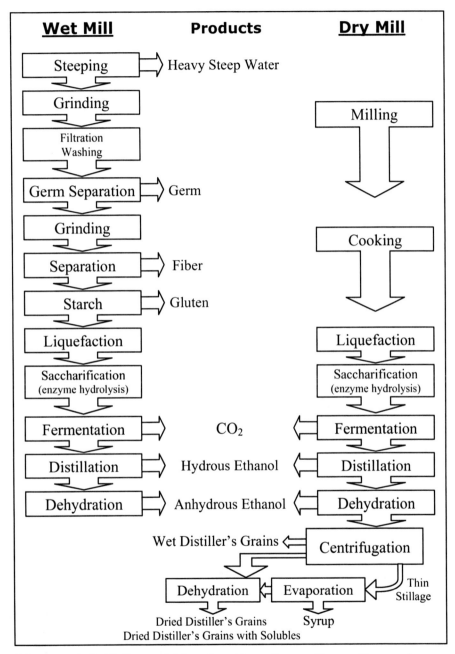

Figure 8-5: Ethanol Production from Corn Kernels
(adapted from *The Dry-Mill Ethanol Industry*, U.S. Department of Energy National Biobased Products and Bioenergy Coordination Office)

Adding Value to Coproducts

When researchers calculate the net energy balance resulting from ethanol production, they must assign an energy value to coproducts. There are various ways to do this. In the case of dry milling, DDGS is used as a livestock feed, so the energy credit might be based on the energy required for the production of traditional feeds that are displaced. New fractionation technologies already in commercial use add more value to coproducts, even with dry milling. The process is known as dry fractionation. The DDGS animal feed resulting from dry fractionation has lower fiber content, making it better for poultry and hogs. Dry fractionation also reduces energy input per gallon of ethanol produced.[28]

FOOD FOR HUMANS? With fractionation, various components of the corn kernel can be marketed as human food products, chemicals, and other valuable materials. Only the low grade starch is diverted to ethanol production. Perhaps this calls for a change in the way we conceive of ethanol's energy balance. Ethanol almost becomes a secondary product. In some cases, most of the energy inputs into growing and milling corn kernels could be counted against the food products and other coproducts rather than the ethanol. We need to eat, after all, and diversion of the low-grade starch does not take much food value from the corn. This concept is illustrated by farmer-owned Lifeline Foods of St. Joseph, Missouri. Lifeline makes corn products for human consumption. In 2007 they commenced operation of an ethanol biorefinery. It makes use of lower quality corn starches coming from the food production facility. The higher quality starches are still used for human consumption, maximizing the efficiency with which corn kernels are utilized.[29]

Some ethanol biorefineries are being designed from the start for simultaneous production of human food and ethanol through dry fractionation. Broin Companies (Now Poet LLC), for instance, has been using dry fractionation on a commercial scale since 2003. They have been able to increase ethanol output and produce valuable coproducts while decreasing energy use.[30]

EnerGenetics Inc. of Iowa is developing a biorefinery concept for adding value to grain while producing ethanol or butanol fuel and other valuable coproducts. A bushel of grain will have a net value as high as $45.00, they say, after processing in one of their biorefineries.[31] The biofuel output from that bushel will comprise only a small fraction of that $45.00 value. It's a whole new model of bioenergy and biomaterial production.

BIODIESEL FROM CORN OIL: Dry-mill biorefineries can now add value to coproducts without installing expensive front-end fractionation equipment. GS CleanTech Corporation sells a relatively inexpensive centrifuge system that will extract corn oil from distiller's grains.[32] The oil can be made into biodiesel. Every bushel of corn can yield about .20 gallons of biodiesel. When combined with the average 2.8 gallons of ethanol production, this results in about 3 gallons of biofuel per bushel of corn.

Ethanol producer VeraSun Energy Corporation is building a 30 million gallon-per-year biodiesel production facility that will use corn oil left over from ethanol production. In a press release, they note the value of DDG's as a livestock feed can be enhanced by removal of oil. Fat content is reduced and proteins are concentrated.[33]

CARBON DIOXIDE FOR ALGAE TO BIOFUELS: According to a press release from GS CleanTech Corporation, about one third of the mass of corn entering a standard dry-mill ethanol biorefinery exits as carbon dioxide (CO_2) at the fermentation stage.[34] Some ethanol biorefineries capture CO_2 for uses such as beverage carbonation. Others simply release the CO_2. GS CleanTech Corporation is developing a system for using algae to capture concentrated CO_2 and convert it into biomass and oxygen with the help of sunshine.[35] Captured oxygen could be used to increase the efficiency of combustion processes at the biorefinery. The algae itself can be converted into more ethanol, biodiesel, or other biofuels through enzymatic processes or gasification followed by catalytic conversion.

Another company, XL Dairy Group Inc., is planning to produce ethanol and biodiesel from algae at their Arizona biorefinery by 2009. They will use dairy manure as a process fuel (more details later in this chapter).[36]

If a CO_2 to biofuel system could be perfected, it would be a step forward in production efficiency, net energy balance, and reduction of greenhouse gas emissions.

CARBON DIOXIDE FOR ENHANCED OIL RECOVERY: CO_2 injection is a proven method for increasing crude oil production from well fields. The U.S. Department of Energy (DOE) is testing the use of CO_2 from ethanol biorefineries for enhanced oil recovery in areas where concentrated CO_2 sources were previously unavailable. Initial tests are underway near Russell, Kansas.

> ...It entails using waste heat from a 15-megawatt natural-gas-fired turbine generator to provide thermal energy for a 25 million gallon-per-year corn ethanol plant. The project then recovers some of the CO_2 that is a byproduct of the fermentation process involved in corn ethanol production and uses it for a CO_2 EOR flood in the Hall-Gurney field in central Kansas... Combined, the benefits from integrating power, ethanol production, enhanced oil recovery, and CO_2 sequestration could total $88 million over 10 years... If the project proves feasible, it could open the door for CO_2 floods throughout Kansas. The potential added incremental oil recovered from such an effort could total as much as 600 million barrels of oil in Kansas alone. As many as 6,000 mature oilfields in the state could be saved from abandonment—not to mention the thousands of jobs created from implementing these projects.[37] (U.S. DOE press release)

Injected CO_2 is expected to largely remain trapped underground where it cannot act as a greenhouse gas.

Ethanol from Field Peas

In 2004, USDA Agricultural Research Service microbiologist Nancy Nichols was contacted by farmers looking for a way to add value to field peas, commonly used as a high-protein animal feed. After removing most of the protein, ethanol could be made from pea starch in existing biorefineries.

Ethanol yield from a bushel of field peas is about 1.7 gallons, compared to 2.8 gallons for corn.[38] Trials by the North Central Research Extension Center at Minot resulted in 46 bushels per acre, on average, over 6 years.[39] Double cropping with wheat or soybeans is possible in many areas because peas mature quickly and thrive with cool temperatures. Two crops from the same piece of ground could maximize profits and productivity. Coproduct animal feeds from field peas would be more valuable than from corn due to a higher level of proteins. Finally, since peas are legumes they do not require expensive, energy-intensive nitrogen fertilizers.

Field peas might be a good ethanol crop in areas too cool and dry for corn production, particularly the northern plains of North America. Starch-based ethanol biorefineries already taking advantage of Canada's wheat crop could use field peas as an additional feedstock with little retrofitting. Legumes such as field peas will improve the soil and provide "free" nitrogen as part of a rotation with traditional grain crops.

Ethanol from Grain Sorghum, Wheat, and Barley

Some biorefineries can easily switch between different grains as feedstocks. Grain sorghum (milo), wheat, and barley are already used as ethanol feedstocks. These grains require minimal irrigation compared to corn. In dry areas, this compensates for lower ethanol yields per acre. It takes energy to pump water, and some aquifers are being pumped faster than nature can refill them.

Grain sorghum, wheat, and barley grow in varied climates and soils, extending the potential for starch-based ethanol production beyond the Corn Belt. Dan Brann, Retired Professor from Virginia Polytechnic Institute and State University, estimates the U.S. Southeast alone could grow 8 million acres of barley—enough to make about 1.2 billion gallons of ethanol per year.[40] This would enhance economies in areas currently left out of substantial ethanol production.

Barley can be double-cropped with soybeans in the Southeast U.S., increasing profitability and output per acre. Soybeans host nitrogen-fixing bacteria, reducing the need for fertilization. Since soybeans are the primary feedstock for U.S. biodiesel, this double-cropping system could produce ethanol and biodiesel from the same ground during the same year, along with all the coproducts of ethanol and biodiesel.

Ethanol output from barley could improve thanks to a new combination of enzymes discovered by researchers at the USDA's Agricultural Research Service.[41]

Reducing Process Fuel Use

Process fuels provide the energy needed for biorefineries. As energy prices increase, producers have greater incentive to build more efficient biorefineries requiring less process fuel per unit of ethanol and other outputs. Natural gas is the primary process fuel today, used mainly for distillation and for drying DDGS for storage and shipping.

Ethanol biorefineries have come a long way in terms of energy efficiency. In a 2006 report, USDA researchers review the history of ethanol production efficiency:

> In the early 1980s, in order to dry alcohol to 99.9 percent, isotropic distillation was used to remove the water. Benzene and cyclohexane (both carcinogenic) were used to remove the water.

Today, molecular sieves are used for dehydrating ethanol. Replacing isotropic distillation with molecular sieves eliminates the use of carcinogenic material, eliminates one distillation process, saves as much as $25,000 per installation, and reduces energy costs by up to 20 percent. There are many energy saving technologies used in corn ethanol plants. Heat exchangers are used widely in ethanol plants to capture excess heat from one process and use it in another process...

Distributed control systems (DCS) were introduced in late 1980s, enabling centralized process monitoring and control. This system allowed process instruments, output to pumps and valves, and controller settings to be driven from a computer console located in a central control room.[42]

Improvements in automated production systems continue to boost ethanol production efficiency. Early in 2007, chemical engineers at Carnegie Mellon University announced they had developed a new multicolumn distillation process together with an energy recovery system and mathematical-optimization techniques. The result could be a reduced need for process steam and thus process fuel. The new system could reduce biorefinery operating costs by 60% and reduce the overall cost of making ethanol by 11%.[43] Since distillation is common to most refining methods, the new process should be applicable to almost any feedstock.

LOW TEMPERATURE HYDROLYSIS: A recent advance specific to starch feedstocks involves saccharification—conversion of starch to glucose (sugars). This is normally done at high temperatures, but low temperature hydrolysis reduces the use of process fuels and water.[44] The new process, also known as raw starch hydrolysis, is already being implemented in corn biorefineries.

FEEDING WET DISTILLER'S GRAIN: Much of the energy used in typical corn ethanol biorefineries goes to drying distiller's grain for storage and shipping. According to Purdue University researchers, cost savings of up to 20% can be realized simply by feeding wet distiller's grain close to the ethanol biorefinery.[45] Ethanol can be produced near existing feedlots or feedlots may be co-located with biorefineries. Missouri's Golden Triangle Energy Cooperative, for instance, sells about 85% of its distiller's grain in the wet form.[46] This not only reduces production costs, but the ethanol from such a biorefinery would boast a better net energy balance because less energy went into the drying process.

HYDRATED ETHANOL: According to the Process Design Center, producing hydrated ethanol costs $.20 less per gallon and uses 45% less energy.[47] Brazil has been making and using hydrated ethanol for decades. Chapter 6 describes new technologies such as ethanol boosting and fuel cells that could increase the efficiency of vehicles running on hydrated ethanol.

Reducing Water Use

Biorefineries use water in the refining process. Fortunately, modern biorefineries include water recycling and water treatment features that largely eliminate wastewater discharge and improve water use efficiency. Some biorefineries can treat municipal wastewater to the point where it can be used in the ethanol production process.[48]

According to the Minnesota Department of Natural Resources, the average Minnesota ethanol biorefinery used 5.8 gallons of water per gallon of ethanol produced in 1998. By 2005, that figure was down to 4.2 gallons.[49] In 2004, water use at Minnesota's ethanol biorefineries accounted for less than 3% of the state's industrial processing water use.[50] Ethanol biorefineries could put a strain on local water supplies in certain cases, however. Reductions in water use must continue alongside growth in ethanol production. Technologies such as raw starch hydrolysis and dry fractionation are making this possible.

In their 2006 report, *Water Use by Ethanol Plants*, The Institute for Agriculture and Trade Policy points out that water is not valued as highly as energy in our economy.[51] The low cost of water means less economic pressure for reducing water use. This may be a case where tax policies, incentives, and regulations will be needed in order to balance the water needs of industry, agriculture, households, and wildlife.

A trend toward cellulosic ethanol production could lessen water use. According to Bill Lee, general manager of the Chippewa Valley Ethanol Company in Minnesota, some thermochemical cellulosic processes will use only about 1 gallon of water per gallon of ethanol produced.[52]

Alternative Process Fuels

With the recent volatility in natural gas prices, some ethanol producers are looking at alternatives such as local biomass or biogas from distiller's grain, livestock manure, and landfills.

BIOMASS: Numerous plant materials come under the "biomass" heading—wood waste, crop residue, grasses, fast-growing trees, paper waste, other solid wastes. Central Minnesota Ethanol Co-op is already using local wood waste as their process fuel, replacing natural gas.[53] Another Minnesota producer, Chippewa Valley Ethanol Company, also plans to use biomass process fuels.[54] Both systems will gasify the biomass. These projects could serve as stepping stones to large-scale cellulosic ethanol production by perfecting biomass collection systems and syngas production from biomass. The thermochemical path to cellulosic production involves transformation of syngas into ethanol.[55]

CSEA Co-Operative Inc. of Canada plans to make ethanol from sweet potato, millet, and other feedstocks. Leftover residue will be anaerobically digested, yielding biogas. The biogas will be sold as a substitute for natural gas or burned to produce electricity. About 40% of the resulting electrical power will be used at the biorefinery. Excess power, perhaps enough to supply 6000 homes, will be sold through the existing grid.[56]

BIOGAS FROM DRIED DISTILLER'S GRAIN: Growth in ethanol production from corn will lead to more dried distiller's grain (DDG). Excess DDG not needed for animal feed could be anaerobically digested. According to biorefinery expert Dr. Martha Schlicher, the resulting biogas could replace at least 70% of the natural gas needed for process heat in biorefineries. Material left over from DDG digestion still has value as a crop fertilizer.[57] Natural gas and DDG prices will largely dictate whether biorefineries decide to implement DDG digestion.

LANDFILL GAS: The decomposition of solid waste in landfills produces a gas consisting of about 50% methane (CH_4), 50% carbon dioxide (CO_2), and a small amount of non-methane organic compounds.[58] According to the U.S. EPA, landfills are the largest U.S. source of human related methane gas emissions, a potent greenhouse gas.[59] In the United States, 423 landfill gas energy projects were operating as of January 2007. The EPA identifies another 570 landfills as good candidates for energy projects, representing 1370 megawatts of potential power.[60] Through the Landfill Methane Outreach Program (LMOP), the U.S. EPA helps industries make use of this largely wasted resource:

> Landfill gas is the natural by-product of the decomposition of solid waste in landfills and is comprised primarily of carbon dioxide and methane. By preventing emissions of methane (a powerful greenhouse gas) through the development of landfill gas energy projects, LMOP helps businesses, states, energy providers, and communities protect the environment and build a sustainable future... Methane is a very potent greenhouse gas that is a key contributor to global climate change (over 21 times stronger than CO_2)... Since the program's inception, LMOP's efforts have reduced landfill methane emissions by nearly 21 million metric tons of carbon equivalent (MMTCE). The greenhouse gas reduction benefits are equivalent to having planted 21.2 million acres of forest or removed the annual emissions from 14.9 million vehicles.[61] (U.S. EPA)

In some cases, biorefineries can be built near landfills. Brooks C&D Landfill, owned by the city of Wichita, Kansas, has been supplying landfill gas to ethanol producer Abengoa Bioenergy (formerly High Plains Corporation) since 1998.[62] A similar partnership is getting underway in Nebraska. L.P. Gill Landfill near Jackson will supply landfill gas to the nearby Souixland Ethanol LLC.[63]

Landfill gas might eventually serve as the actual feedstock from which ethanol is made. This would further reduce the fossil fuel input for ethanol production. Syntec Biofuel Research Inc. lists landfill gas as a possible feedstock for their proprietary gas-to-ethanol catalyst technology.[64]

COAL: A few producers use coal as a process fuel. Burning coal releases more greenhouse gases compared to natural gas (unless CO_2 is sequestered), but is potentially better in terms of energy security. North America's natural gas supplies are running low, but coal reserves are relatively plentiful. Clean coal technologies and tightly regulated emissions will be important for utilizing these reserves. It would make no sense to produce clean-burning biofuels at the expense of excessive pollution from process fuels. The design of some coal-fired boilers allow easy conversion for burning biomass such as solid waste or wood chips in the event these materials become more available or more cost effective.[65]

LIGNIN: With biochemical cellulosic ethanol production, leftover lignin could be burned as a superior process fuel. This would result in a cleaner burn, less greenhouse gas emissions, and reduced air pollution (SOx, NOx, and particulate matter) as compared to coal.[66] Lignin could be burned in the biorefinery itself or replace part of the coal in a co-located power plant.

WASTE HEAT: Coal can produce electricity and provide process heat for ethanol production at the same time. This is because coal-fired power generation creates heat which must somehow be dissipated. Missouri's Thomas Hill Energy Center, for instance, is situated near a sizeable man-made lake for cooling purposes. The lake is warmer near the power plant. Co-located biorefineries can make use of this low-grade heat. Co-location can increase the profitability and decrease the environmental impact of both ethanol production and electric power generation. Co-location practically eliminates process fuels from energy balance calculations since heat is a by-product of power generation. Why let the heat go to waste?

Great River Energy (A non-profit electric cooperative) and Headwaters, Inc. joined forces to build an ethanol biorefinery next to an existing coal-fired power station near Underwood, North Dakota. Blue Flint Ethanol LLC does not require boilers or associated infrastructure because it uses waste heat from the Coal Creek Power Station.[67] Ethanol production at Blue Flint began in 2007.[68]

Combined Heat and Power

If a biorefinery can't be located near an existing power plant, it can still incorporate combined electrical power and steam production. A biorefinery can generate its own electricity from coal, biomass, natural gas, biogas, methane, propane, or other fuels, and then capture the resulting low-grade heat for the refining process. It's like getting two energy streams for the price of one. These systems are known as cogeneration or combined heat and power (CHP). Initial construction costs are higher, but operating costs are lower. A 2006 EPA study found a 12.2% decrease in process fuel use when using CHP with natural gas for a typical dry-mill ethanol biorefinery.[69]

U.S. Energy Partners LLC does even better by supplying excess electricity to the city of Russell, Kansas. Their CHP system, installed in partnership with the city, reduces fuel use at the biorefinery by 28% compared to a typical system. The CHP turbines produce 15 megawatts of electricity, 3 of which are used in the biorefinery. Turbine exhaust is used to produce steam needed for ethanol production.[70] Northeast Missouri Grain LLC enjoys a similar partnership with the city of Macon, Missouri. Their process steam and electrical power come from a 10 megawatt natural gas-fired turbine.[71]

Adkins Energy LCC, a farmer-owned ethanol biorefinery in Northern Illinois, generates about 99% of its electrical needs from an on-site 5 megawatt turbine. A heat recovery system recycles waste heat from the turbine, providing about 32% of the thermal needs for the ethanol production process. With a natural gas cost of $5.50 per million BTU's, the Adkins Energy CHP saves approximately $903,000 per year. The expected payback time on the initial investment is around 3.3 year.

Another advantage is elimination of shut-downs due to power outages. The biorefinery is connected to the grid for back-up purposes, but can operate without grid power when necessary.[72]

Central Minnesota Ethanol Co-op is taking CHP a step further by using wood waste as the process fuel. The wood biomass is gasified, resulting in a "producer gas" that powers the CHP unit. According to an article in *Ethanol Producer Magazine*, this CHP system enables the production of ethanol with an energy balance of 3.15:1 (1 unit of fossil energy required for every 3.15 units of energy available from the ethanol), while the industry average is only about 1.67:1. The CHP system began operation in 2006. The gasifier was built by Primenergy LLC of Tulsa, Oklahoma. Sebesta Blomberg handled biorefinery integration.[73] This type of system could be powered by various biomass sources, taking advantage of local availability.

Other ethanol biorefineries with CHP systems include East Kansas Agri Ethanol of Garnett, Kansas and Otter Creek Ethanol of Ashton, Iowa.[74]

Advantages of Combined Heat and Power (CHP) for Ethanol Production[75]

- CHP can improve the economics of ethanol production in areas with high electric rates, yielding energy savings of 10 to 25 percent.
- On-site power generation with CHP can provide a hedge against unstable electric rates and unreliable electric resources.
- CHP can ensure that a plant keeps operating, even when the surrounding electric grid is down.
- CHP can offer the opportunity to partner with your municipal utility or rural cooperative to leverage resources.
- CHP systems can be designed to operate on any fuel, ensuring that your plant optimizes the use of available energy resources, improving its overall efficiency and competitive position in the marketplace.

- Compared to conventional systems, CHP greatly reduces total energy use and the resulting emissions of carbon dioxide (CO_2), a contributor to global climate change.

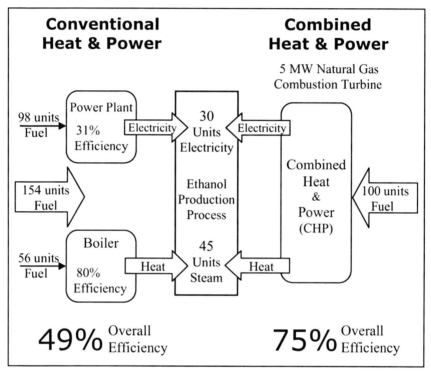

Figure 8-6: Conventional vs. Combined Heat & Power
(adapted from the U.S. EPA, *Combined Heat and Power*)[76]

Ethanol-Livestock Integration

Some ethanol producers are combining many of the technologies discussed above into super-efficient systems coined "closed loop." Each technology is already proven in commercial use. Many waste streams are recycled, creating partially self-sustaining systems. This results in a better ethanol energy balance and less environmental damage.

E³ BIOFUELS LLC: A "closed loop" system is already operating at Mead, Nebraska.[77] E³ Biofuels LLC built their "Genesis" ethanol biorefinery near an existing cattle feedlot equipped with slatted floors and a manure collection system. Biogas made by anaerobic (oxygen-free) digestion of manure and corn stalks is used in place of natural gas for process heat at the biorefinery. An aqueous ammonia fertilizer is produced from the manure as a coproduct of the process. It can be injected below the soil surface, avoiding much of the nutrient loss associated with traditional manure applications. Corn kernels are processed in the usual way. The wet distiller's grain by-product is fed to cattle, closing the loop. Feeding distiller's grain on-site saves energy that would normally be expended on transportation and drying (drying prevents spoilage in transit).

E^3 biofuels estimates their system achieves a 5:1 "corn-to-consumer" energy balance, in contrast to the 2004 USDA estimate of 1.67:1 for a standard corn ethanol biorefinery. This improvement results from combining several features including:

- on-site cattle feeding with manure collection system
- Biogas derived from manure and cornstalks is used as a process fuel in place of natural gas.
- fertilizer produced from manure after biogas removal
- Feeding wet distiller's grain saves energy normally used in drying.
- Feeding distiller's grain on-site saves energy normally used for transportation.
- No distiller's grain drying system is needed.

E^3 Biofuels plans to build additional "closed loop" biorefineries across the Midwest.

PANDA ETHANOL: Another manure-powered ethanol biorefinery is under construction near the Texas panhandle town of Hereford. Each one of the 3.5 million cattle located within 100 miles of Hereford produces about 1000 pounds of manure each year.[78] Rather than paying to have it carted away, feedlots will provide up to 1 billion pounds of manure per year to Panda Ethanol Inc. at no cost. Manure and cotton gin waste will be converted to biogas with a bubbling bed fluidized gasifier.[79] It is estimated Panda will save the equivalent of 1,000 barrels of oil per day by using biogas as their process fuel. They also plan to make use of gray water from a city wastewater facility.[80]

At 100 million gallons-per-year of ethanol output, the Hereford facility will be the largest biomass-fueled ethanol biorefinery in the United States. Panda Ethanol Inc. plans to build more such biorefineries in the near future.[81]

THE XL DAIRY GROUP INC: In Vicksburg, Arizona, the XL Dairy Group Inc. is building an integrated biorefinery including a dairy, fractionation mill (corn milling), ethanol plant, and biodiesel plant. The company expects to achieve a 10:1 net energy ratio for ethanol by generating power on-site from waste products and utilizing all coproducts. The integration is expected to reduce overall emissions and dairy odors as well. Full operations are expected by the fall of 2008. Future plans include biofuel production from algae with the help of carbon dioxide from the biorefinery and "fertility water" from the methane digesters. The proprietary algae-to-biofuel process is expected to be deployed on a commercial scale by 2009.[82]

Small-Scale Ethanol Production

Just because industrial-scale ethanol production has become successful does not mean we should abandon the idea of making fuel on the farm to be used on the farm or in the local community. In the past, farmers raised their own transportation fuel in the form of hay and grain for horses—now they can do the same in the form of farm-produced biofuels. Many books and web sites feature ethanol recipes and instructions. Small stills are available for sale as well.

Feedstocks with readily available simple sugars generally require fewer steps and less effort for conversion to ethanol. These include sugar cane, sweet sorghum, sugar beets, Jerusalem artichokes tops (before blossoming), and waste fruit. Sugar beets and Jerusalem artichokes are fairly easy to grow on a small scale in most regions of North America. Sweet sorghum thrives throughout much of North America, while sugar cane thrives in tropical and sub-tropical climates. Home-scale ethanol production is feasible for starchy feedstocks like corn kernels, cattails and Jerusalem artichoke roots as well.

Practically all water must be removed for the production of anhydrous ethanol. This is possible on a small scale, but more difficult than producing a lower proof ethanol with a little more water in it (known as hydrated or hydrous ethanol). One approach used by ethanol enthusiasts has been converting engines to run on hydrated ethanol.

For ethanol production on any scale, a permit from the U.S. Tobacco Tax and Trade Bureau (TTB) is required by law. Their web site is:

www.ttb.gov/tax_audit/permits.shtml

Additional licenses or permits may be required at the federal or state level. Requirements are generally more stringent if the ethanol will be sold. Your state department of agriculture or local county extension office should be able to help with any necessary permits, taxes, or licenses.

Ethanol Transportation & Pipeline Issues

Mostly because of ethanol's affinity for water, it presents challenges for shipping in pipelines also used for petroleum products. It is not impossible, however. In the early 1980's, the Williams Company successfully shipped 4,600 barrels of neat (without gasoline or other additives) ethanol from Kansas City to Des Moines in a pipeline used for multiple products.[83] Based on their experience, the company developed recommended procedures for routine ethanol shipment in multi-product pipelines. These procedures include frequent mainline dewatering. "…our experimental pipeline tests," said a Williams representative, "indicate that fuel grade ethanol can be successfully transported in a multi-products pipeline system under controlled conditions."[84]

Today, North American ethanol is shipped by barge, rail, or truck. Construction of biorefineries in multiple regions of North America will reduce distance to market. Production will move closer to the end user as cellulosic biorefineries begin using feedstocks prevalent outside the Corn Belt. Still, pipelines might eventually make sense for supplying population centers.

Dedicated ethanol pipelines are possible, but North American production levels have not yet warranted this option. In 2006, Senator Tom Harkin (D-IA) and Senator Richard Lugar (R-IN) proposed a study looking into the feasibility of an ethanol pipeline system. "With the ethanol production and demand both on the rise," said Harkin, "we need an accurate and fair analysis of the potential to distribute ethanol around the country by pipeline. We continue to hear comments that it can't be done efficiently, but it is happening right now in Brazil. The goal of this bill is to examine the issue and get all the facts on the table."[85]

In 2008, construction is scheduled to begin on a "multi-pipeline" in Brazil. It will have the capacity to move 18 million gallons per year of ethanol, gasoline, or diesel to the Port of Paranaguá.[86]

Switching to butanol—a different type of alcohol—could be another strategy for easier transportation and distribution. Compared to ethanol, butanol is more easily transported through existing pipelines also used for petroleum products.

Butanol: The Other Alcohol

If we could wave a magic wand and create a better biofuel, it might look like butanol (also known as butyl alcohol). Butanol is an alcohol like ethanol, but with a 4-carbon structure and the molecular formula $C_4H_{10}O$ (Ethanol is a 2-carbon alcohol).[87] Butanol is sometimes referred to as "biobutanol" when made from renewable feedstocks such as grain, sugar, or cellulosic material, as opposed to fossil fuels.

The U.S. National Renewable Energy Laboratory lists several advantages of butanol as a fuel:[88]

- much higher volumetric energy content than ethanol
- does not suffer from separation caused by water
- Gasoline-butanol blends appear more compatible with pipeline system—needs to be verified.
- Gasoline containing butanol (up to 2.7% oxygen) is already "approved" by EPA.
- may not suffer from non-ideal vapor pressure (vapor pressure bump) like ethanol
- may lower vapor pressure of ethanol blends

Butanol contains more BTU's per gallon (93,000) as compared to ethanol (76,000).[89] This results in a fuel economy closer to that of gasoline, even without engine optimization. If engines were designed for butanol's higher octane, an even better fuel economy could be achieved. Compared to ethanol, butanol can be blended with gasoline at a higher percentage for use in normal (non-flex) automobiles. This could mean greater market penetration without the need for more flex-fuel vehicles.

With all these advantages, why aren't we making butanol instead of ethanol? Up until now, production costs have simply been too high. That could be changing. ButylFuel™ LLC of Ohio says they have developed a new butanol production method that results in 2.5 gallons of butanol from a bushel of corn, with hydrogen as a coproduct.[90] An Iowa Company, EnerGenetics International Inc., is working on butanol production with nutraceuticals and bioplastics as possible coproducts in addition to hydrogen. Their "Mini-Biorefineries" will use feedstocks such as corn kernels or sweet sorghum cane. According to a press release, the company demonstrated butanol recovery from grape and citrus pomace and sweet sorghum cane at the pilot plant level.[91] In 2006, Dupont and BP announced a partnership to "develop, produce and market a next generation of biofuels."[92] Their first product to market will be biobutanol for blending with gasoline in the United Kingdom. They expect to begin production by the end of 2007. If successful, these butanol initiatives could launch a new era in biofuels, while building on the success of the ethanol industry.

Butanol can be made from the same feedstocks as ethanol, and in the same biorefineries, with some retrofitting. A BP press release notes biobutanol can even enhance the performance of ethanol blends by "reducing ethanol's impact on vapour pressure."[93] When ethanol is blended with gasoline at low levels (usually E10), it can force the use of special gasoline blendstock or on-board vapor recovery systems to counteract the higher vapor pressure.[94] Butanol added to the blend along with ethanol could moderate vapor pressure, reducing the need for these measures.

Technologies already in use prove corn ethanol and other biofuels can be produced more sustainably than in the past. The biofuel industry can do even more toward sustainability while also boosting output. The next step will be an expansion beyond corn with cellulosic ethanol and butanol.

Notes

1. USDA Energy Efficiency and Renewable Energy Biomass Program, *Integrated Biorefineries,* 2005, http://www1.eere.energy.gov/biomass/integrated_biorefineries.html.
2. Vinod Khosla of *Khosla Ventures* (http://www.khoslaventures.com/) is a major investor in both grain-based and cellulosic ethanol ventures. He was a keynote speaker at the 2006 American Coalition for Ethanol annual convention in Kansas City and at other events across the nation.
3. Milton Maciel, "Ethanol from Brazil and the USA," *ASPO-USA /Energy Bulletin,* October 2006, http://www.energybulletin.net/21064.html.
4. Fernando Reinach, "Biofuels in Brazil: Today and in the Future," Presentation at the *4th annual Life Sciences & Society Symposium* at the University of Missouri, Columbia, March 15, 2007.
5. Hosein Shapouri and Michael Salassi, "The Economic Feasibility of Ethanol Production from Sugar in the United States," *U.S. Department of Agriculture*, July 2006, http://www.usda.gov/oce/EthanolSugarFeasibilityReport3.pdf.
6. Gwen Roland, "Jackie Judice and family, Northside Planting LLC, Franklin, Louisiana," in *The New American Farmer, 2nd Edition,* (Sustainable Agriculture Research and Education, 2005), http://www.sare.org/publications/naf2/judice.htm

7. Y. Terajima and A. Sugimotoby, *Breeding higher yield sugarcane which are adaptable to adverse condition as dry and low fertility*, http://unit.aist.go.jp/internat/biomassws/03workshop/material/sugimoto.pdf.
8. Aya Takada, "Japan Brewer Pursues 'Monster Cane' Ethanol Dream," *Reuters*, October 18, 2006, http://www.planetark.com/dailynewsstory.cfm?newsid=38535&newsdate=18-Oct-2006.
9. EnerGenetics International Inc., http://www.energeneticsusa.com.
10. I.C. Anderson, "Ethanol from Sweet Sorghum," *Iowa State University*, http://www.energy.iastate.edu/renewable/biomass/cs-anerobic2.html.
11. Sorganol Production Co. Inc., http://www.sorganol.com.
12. Oklahoma State University, "Converting Sweet Sorghum into Ethanol," news release, March 19, 2006. http://osu.okstate.edu/index.php?option=com_content&task=view&id=7&Itemid=90.
13. Abengoa Bioenergy, *Energy Crops*.
14. L. Baker, P.J. Thomassin, J.C. Henning, "The Economic Competitiveness of Jerusalem Artichoke (Helianthus tuberosus) as an Agricultural Feedstock for Ethanol Production for Transportation Fuels," *Canadian Journal of Agricultural Economics* 38 (1990), 981–990.
15. Tony Peluso, Laurie Baker, & Paul J. Thomassin, "The Siting of Ethanol Plants in Quebec," *Canadian Journal of Regional Science,* 21 (1998).
16. W.C. Fairbank, "Jerusalem Artichoke for Ethanol Production," *University of California University Extension/Vegetable Research and Information Center,* http://vric.ucdavis.edu/veginfo/commodity/jerusalem%20artichoke/Engineersreportartichoke.pdf
17. Ethanol Producer Magazine, *Wisconsin developer provides cheese industry solution*, June 2006, http://ethanolproducer.com/article.jsp?article_id=2028.
18. Renewable Fuels Association, *Ethanol biorefinery Locations*, March 2007, http://www.ethanolrfa.org/industry/locations/.
19. Ontario Power Authority, *Biomass Energy Case Study 02: CSEA Cooperative Inc.*, 2006.
20. Tyler Hamilton, "Sweet Dreams," *Toronto Star*, January 15, 2007, http://www.thestar.com/article/171058.
21. Florida Fruit & Vegetable Association, "Watermelon Watershed," *Harvester Online*, September 2006, http://www.ffva.com/publications/harvester/sep06_ETH.asp.
22. Daniel West, "Ethanol From Waste Fruit," *Sustainable Agriculture Research and Education*, 2006, http://www.sare.org/reporting/report_viewer.asp?pn=FNC03-464&ry=2006&rf=1
23. Ibid.
24. Hosein Shapouri and Michael Salassi, "The Economic Feasibility."
25. U.S. Department of Energy Office of Science, *Fuel Ethanol Production,* 2006, http://genomicsgtl.energy.gov/biofuels/ethanolproduction.shtml.
26. U.S. Department of Energy Office of Science, *Fuel Ethanol Production.*

27. Martha Schlicher, "Biofuels in the U.S.: Today and in the Future," Presentation at the *4th annual Life Sciences & Society Symposium* at the University of Missouri, Columbia, March 15, 2007.
28. Hosein Shapouri and Michael Salassi, "The Economic Feasibility."
29. Missouri Corn Growers Association, "Missouri's Newest Ethanol Plant Produces Both Food & Fuel," news release, August 24, 2007, http://www.mocorn.org/.
30. Country Journal Publishing Co., *Broin Companies Announces Fractionation Process for Ethanol Production*, July 5, 2006, http://www.grainnet.com/articles/Broin%20Companies%20Announces%20Fractionation%20Process%20for%20Ethanol%20Production-26898.html.
31. EnerGenetics International Inc., *Company Description*, http://www.energeneticsusa.com/.
32. GS CleanTech Corporation, "GS CleanTech to Present at the Iowa Renewable Fuels Summit During 'The Future is Now' Panel Review," news release, February 28, 2007, http://www.gs-agrifuels.com/news.php?id=39.
33. VeraSun Energy Corporation, "VeraSun Announces Innovative Process for Biodiesel Production," news release, November 3, 2006, http://www.verasun.com/.
34. GS CleanTech Corporation, "GS CleanTech Releases Process Demonstration of CO2 Bioreactor Technology," news release, September 6, 2006, http://www.greenshift.com/news.php?id=181.
35. Ibid.
36. XL Dairy Group, "XL Biorefinery–Vicksburg completes Phase I dairy—Vicksburg, Arizona," news release, April 27, 2007, http://xldairygroup.com/.
37. DOE National Energy Technology Laboratory, *CO2 Injection Boosts Oil Recovery, Captures Emissions*, January 3, 2005, http://www.netl.doe.gov/publications/press/2005/tl_kansas_co2.html.
38. Jan Suszkiw, "Imagine—Fuel Alcohol From Pea Starch!," *Agricultural Research Magazine*, March 28, 2006, http://www.ars.usda.gov/is/pr/2006/060328.htm.
39. Kent McKay, Blaine Schatz, and Gregory Endres, "Field Pea Production," *North Dakota State University Agriculture and University Extension*, March 2003, http://www.ag.ndsu.edu/pubs/plantsci/rowcrops/a1166w.htm.
40. Kevin B. Hicks et. al., "Current and Potential Use of Barley in Fuel Ethanol Production," *2005 EWW/SSGW Conference*, May 9–12, Bowling Green, KY, http://www.uky.edu/Ag/Wheat/wheat_breeding/EWW_SSGW/documents/kevin_hicks.doc
41. USDA Agricultural Research Service, *Economic Competitiveness of Renewable Fuels Derived from Grains and Related Biomass 2006 Annual Report*, 2007.
42. Hosein Shapouri and Michael Salassi, "The Economic Feasibility."
43. Chriss Swaney, "Engineers Devise Process to Improve Energy Efficiency of Ethanol Production," *Carnegie Mellon University*, January 26, 2007, http://www.cmu.edu/news/archive/2007/January/jan26_ethanol.shtml.

44. Hosein Shapouri and Michael Salassi, "The Economic Feasibility."
45. Nathan S. Mosier and Klein Ileleji, "How Fuel Ethanol Is Made from Corn," *Purdue Extension,* 2006, http://www.ces.purdue.edu/extmedia/ID/ID-328.pdf.
46. Golden Triangle Energy Cooperative, http://www.goldentriangleenergy.com/feed1.htm
47. Process Design Center, *Ethanol in Gasoline –Hydrous E15*, presented at *Conference Sustainable Mobility,* November 21, 2006, http://www.energyvalley.nl/uploads/Mr_Keuken.pdf.
48. Dennis Keeney and Mark Muller, "Water Use by Ethanol Plants," *Institute for Agriculture and Trade Policy,* October 2006, 4.
49. Ibid.
50. MDNR Water Appropriations Permit Program, 2004, http://www.pca.state.mn.us/publications/presentations/ethanol-0706-manning.pdf
51. Keeney and Muller, "Water Use by Ethanol Plants."
52. Andrea Johnson, "Farmers ask questions about cellulosic ethanol production," *Farm and Ranch Guide,* February 16, 2007, http://www.farmandranchguide.com/
53. Ethanol Producer Magazine Staff Report, "CMEC Gasifier Starts Production," December 2006.
54. Gretchen Schlosser, "Gasifier project awaits MPCA permit, designs being finalized," *West Central Tribune,* January 23, 2007.
55. USDA Energy Efficiency and Renewable Energy Biomass Program, *Integrated Biorefineries,* 2005, http://www1.eere.energy.gov/biomass/integrated_biorefineries.html.
56. Ontario Power Authority, *Biomass Energy Case Study 02: CSEA Cooperative Inc.,* 2006, http://www.powerauthority.on.ca/Page.asp?PageID=122&ContentID=4009&SiteNodeID=253.
57. Martha Schlicher, "Biofuels in the U.S.: Today and in the Future."
58. U.S. EPA Landfill Methane Outreach Program, http://www.epa.gov/lmop/overview.htm
59. U.S. EPA, *Inventory of U.S. Greenhouse Gas Emissions and Sinks: 1990–2004,* April 2006, 8–1.
60. U.S. EPA, *Energy Projects and Candidate Landfills,* http://www.epa.gov/lmop/proj/index.htm
61. U.S. EPA Landfill Methane Outreach Program, http://www.epa.gov/lmop/.
62. City of Wichita, *Brooks Construction and Demolition Waste Landfill,* http://www.wichita.gov/CityOffices/PublicWorks/Brooks/.
63. Nate Jenkins, "Methane greens operation of ethanol plants," *Associated Press,* January, 28, 2007, http://www.insidebayarea.com/business/ci_5105873; Ethanol Producer Magazine, "Nebraska's Siouxland Ethanol Breaks Ground," December 2005, http://www.ethanolproducer.com/article.jsp?article_id=242.
64. Syntec Biofuel, *Technology,* 2004, http://www.syntecbiofuel.com/technology.html

65. Energy Products of Idaho describes their clean coal technology and biomass-capable boiler system at http://www.energyproducts.com/goldfield.htm.
66. Bob Wallace, Mark Yancey, and James Easterly, "Co-locating a Cellulosic Biomass to Ethanol Plant with Existing Coal Fired Power Plants," *U.S. Department of Energy Office of Energy Efficiency and Renewable Energy*, http://www1.eere.energy.gov/biomass/pdfs/34534.pdf.
67. Blue Flint Ethanol, *Ethanol Production Using Waste Heat*, January 29, 2007, http://www.blueflintethanol.com/data/upfiles/project/UsingWasteHeat.pdf.
68. Headwaters Incorporated, "Headwaters Incorporated Announces Start-up of the Blue Flint Ethanol Facility," news release, February 21, 2007.
69. Energy and Environmental Analysis, Inc. for the U.S. EPA Combined Heat and Power Partnership, *An Assessment of the Potential for Energy Savings in Dry Mill Ethanol Plants from the Use of Combined Heat and Power (CHP)*, July 2006, http://www.epa.gov/chp/pdf/ethanol_energy_assessment.pdf.
70. U.S. EPA Combined Heat and Power Partnership, *Combined Heat and Power*, http://www.epa.gov/chp/pdf/ethanol_factsheet.pdf.
71. U.S. EPA Combined Heat and Power Partnership, *Combined Heat and Power;* also see Northeast Missouri Grain LLC, http://nemog.aghost.net/index.cfm.
72. U.S. EPA Combined Heat and Power Partnership, *Adkins Energy LLC 5 MW CHP Application*, http://www.epa.gov/CHP/pdf/Adkins%20Energy.pdf.
73. Ethanol Producer Magazine Staff Report, "CMEC Gasifier Starts Production," December 2006.
74. U.S. EPA Combined Heat and Power Partnership, *Combined Heat and Power*.
75. Ibid.
76. Ibid.
77. E3 Biofuels, http://www.e3biofuels.com/.
78. Panda Ethanol, "Panda Ethanol to Break Ground on 100 Million Gallon Ethanol Plant in Hereford, Texas," news release, September 14, 2006, http://www.pandaethanol.com/portals/0/pdf_files/Hereford_Groundbreaking_Release_091406.pdf.
79. Texas State Energy Conservation Office, *Biomass Energy: Manure for Fuel*, 2006, http://www.seco.cpa.state.tx.us/re_biomass-manure.htm.
80. Ibid.
81. Panda Ethanol, *Panda Ethanol to Break Ground*.
82. XL Dairy Group, "XL Biorefinery–Vicksburg completes Phase I dairy.—Vicksburg, Arizona," news release, April 27, 2007, http://xldairygroup.com/.
83. Ron Miller, "Production of Ethanol & Update on Ethanol Current Events," *Williams Bio-Energy*, April 10, 2001, http://www-erd.llnl.gov/ethanol/proceed/etohupd.pdf.
84. Williams Presentation, Alcohol Week Conference (San Antonio), March 1982, Reproduced in Ron Miller, "Production of Ethanol & Update on Ethanol Current Events," *Williams Bio-Energy*, April 10, 2001, http://www-erd.llnl.gov/ethanol/proceed/etohupd.pdf

85. Office of Senator Tom Harkin, "Harkin Calls for Study of Pipeline Distribution of Ethanol," news release, September 29, 2006, http://harkin.senate.gov/news.cfm?id=264171.
86. Omar Nasser, "Multi-pipeline will take ethanol from Mato Grosso to the Port of Paranaguá," *Brazil-Arab News Agency,* March 2, 2007, http://www.anba.com.br/ingles/noticia.php?id=13964.
87. U.S. Environmental Protection Agency, "Chemical Summary for 1-Butanol," August 1994, http://www.epa.gov/chemfact/s_butano.txt.
88. Robert L. McCormick, "Liquid Fuels from Biomas," *U.S. DOE National Renewable Energy Laboratory*, August 23, 2006, http://www1.eere.energy.gov/vehiclesandfuels/pdfs/deer_2006/plenary3/2006_deer_mccormick.pdf.
89. Ibid.
90. Environmental Energy Inc., http://www.butanol.com/.
91. EnerGenetics International Inc., "The Next Generation In Biofuels: Bio-Based Butanol," news release, November 2006.
92. BP, "DuPont Bio-based Science and BP Fuels Technology Expertise Will Bring Next Generation Biofuels to Market," news release, June 20, 2006, http://www.bp.com/; Also see FAQ's at http://www2.dupont.com/Biofuels/en_US/FAQ.html
93. Ibid.
94. Marika Tatsutani, ed., "Health, Environmental, and Economic Impacts of Adding Ethanol to Gasoline in the Northeast States, Volume 2, Air Quality, Health, and Economic Impacts," *Northeast States for Coordinated Air Use Management*, 2001, 14.

Chapter 9

CELLULOSIC ETHANOL

Many supporters of renewable biofuels see cellulosic ethanol as our best hope in the next few decades. The attraction lies in the sheer volume of potential feedstocks. Cellulosic ethanol (or butanol) does not refer to a distinct end product. Ethanol made from cellulosic materials is the same as ethanol from corn or sugar cane. The distinction lies in the feedstocks. Cellulosic biofuels can be made from practically any organic material. The U.S. Department of Energy (DOE) Biomass Program lists the following categories of cellulosic feedstocks:[1]

- agricultural residues (leftover material from crops, such as the stalks, leaves, and husks of corn plants)
- forestry wastes (chips and sawdust from lumber mills, dead trees, and tree branches)
- municipal solid waste (household garbage and paper products)
- food processing and other industrial wastes (black liquor, a paper manufacturing by-product)
- energy crops (fast-growing trees and grasses) developed just for this purpose

Diverse feedstocks will allow an expansion in ethanol output. The challenge is in liberating plant sugars from the grip of cellulose, hemicellulose, and lignin. The following descriptions of these complex polymers are from the U.S. DOE:

- Cellulose is the most common form of carbon in biomass, accounting for 40%–60% by weight of the biomass, depending on the biomass source. It is a complex sugar polymer, or polysaccharide, made from the six-carbon sugar, glucose. Its crystalline structure makes it resistant to hydrolysis, the chemical reaction that releases simple, fermentable sugars from a polysaccharide.

- Hemicellulose is also a major source of carbon in biomass, at levels of between 20% and 40% by weight. It is a complex polysaccharide made from a variety of five and six-carbon sugars. It is relatively easy to hydrolyze into simple sugars but the sugars are difficult to ferment to ethanol.
- Lignin is a complex polymer, which provides structural integrity in plants. It makes up 10% to 24% by weight of biomass. It remains as residual material after the sugars in the biomass have been converted to ethanol. It contains a lot of energy and can be burned to produce steam and electricity for the biomass-to-ethanol process.[2]

Commercializing Cellulosic Production

Is cellulosic ethanol ready for prime time? In the 1970's, researchers at the U.S. Army's Natick Development Center believed conversion of cellulose to glucose (sugars) would be "technically feasible and practically achievable on a very large scale by 1980."[3] It didn't happen that quickly. Petroleum prices plummeted and cellulosic conversion research went largely unfunded until the new millennium. Technology for cellulosic ethanol exists, but costs are high thanks to plant cell structures resistant to disintegration into simple sugars.

Plants use three main materials to build their cell walls: the polysaccharides cellulose and hemicellulose and the phenolic polymer lignin. Cellulose is a chain of glucose (sugar) molecules strung together. As these molecules multiply, they organize themselves in linear bundles that crisscross through the cell wall, giving the plant strength and structure.

The cellulose bundles are weakly bound to an encircling matrix of hemicellulose, which is strongly linked to lignin. The gluey lignin polymer further strengthens plants and gives them flexibility. Lignin is the reason plants can pop back up after heavy rains and winds. And it's how they made the leap from a life in the ocean to one on land eons ago.

Plants have invested great energy in crafting exquisite cell-wall structures that resist degradation and loss of their precious sugars. Over the course of millions of years, they've had to fend off an insatiable crowd of energy-hungry fungi, bacteria, herbivores—and now, people.[4] (USDA ARS)

We are finally on the verge of a cellulosic revolution because of sustained high energy prices, private investment, and a government jump-start. On February 28, 2007, The U.S. Department of Energy announced up to $385 million in Federal funding for six potential cellulose-to-ethanol biorefinery projects over a four-year period. These biorefineries could produce more than 130 million gallons of ethanol per year when fully operational.[5] The DOE released the following information about the six projects selected for possible funding:[6]

Abengoa Bioenergy Biomass of Kansas LLC

Corporate HQ: Chesterfield, Missouri
Proposed Facility Location: Colwich, Kansas
Description: This project from a committed long-term player has the potential to demonstrate dual biochemical and thermochemical capabilities.
Participants: Abengoa Bioenergy R&D, Abengoa Engineering, Antares Corp., Taylor Engineering
Production:
- 11.4 million gallons/year and sufficient energy to power the operation and sell excess energy to the co-located dry-grind ethanol production plant
- Both ethanol and syngas production, with long term strategy of using the syngas for ethanol and chemicals production

Technology & Feedstocks:
- Thermochemical and biochemical processing of 700 tons/day of corn stover, wheat straw, milo (sorghum) stubble, switchgrass, and other opportunity feedstocks

State of Readiness:
- Abengoa is building a 1.2 ton/day facility at York, NE to evaluate an integrated bioprocess under a current DOE award and a

70 ton/day integrated process for both the bio and thermochemical processes in Spain.
- Both plants are under construction and will be starting operations in 2007.
- Integrated commercial plant construction starts in late 2008 and planned project completion is in late CY 2011.

ALICO Inc.

Corporate HQ: LaBelle, Hendry County, Florida
Proposed Facility Location: LaBelle, Florida
Description: This is a major agricultural company that will use an innovative thermochemical technology that ferments synthesis gas. This thermochemical/fermentation conversion alternative could be suitable to feedstocks that are not easily processed biologically such as wood and wood wastes.
Participants: Bioengineering Resources, Inc. of Fayetteville, Arkansas; Washington Group International of Boise, Idaho; GeoSyntec Consultants of Boca Raton, Florida; BG Katz Companies/JAKS,LLC of Parkland, Florida; Emmaus Foundation, Inc.
Production:
- Ethanol, electricity, ammonia and hydrogen
- 7 million gallons/year from first unit; second unit at 13.9 million gallons/year, 6,255 KW power plus ammonia and hydrogen (produced based on market requirements including their own use of ammonia as fertilizer).

Technology & Feedstocks:
- Gasification of wood or agricultural residues and fermentation of syngas to ethanol with ammonia and electricity as coproducts
- 770 tons/day of yard, wood and vegetative wastes (citrus peel); eventually energycane

State of Readiness:
- Demonstrated an integrated process at pilot small scale for 6 years
- Construction Start: CY 2008; Project completion: late CY 2010

BlueFire Ethanol Inc.

Corporate HQ: Irvine, Orange County, California

Proposed Facility Location: Southern California
Description: This company has experience building biomass power plants in California and their technology has been demonstrated at the pilot scale. One of its partners is Waste Management Inc., a leading waste-to-energy company. This project will give DOE understanding of a new biological fermentation process not using enzymes.
Participants: Waste Management, Inc., JGC Corporation; MECS Inc.; NAES; PetroDiamond
Production:
- 19 million gallons/year in the unit in which DOE will be participating.

Technology & Feedstocks:
- Concentrated acid processing of 700 tons/day of sorted green waste and wood waste from landfills followed by fermentation of sugars to ethanol

State of Readiness:
- Blue Fire and their Japanese partner have demonstrated an integrated process for a few years at a smaller scale in Japan.
- Construction Start: 2008; Project completion: end of CY 2009.

Poet LLC (formerly Broin Companies)

Corporate HQ: Sioux Falls, South Dakota
Facility Location: Emmetsburg, Palo Alto County, Iowa
Description: This Midwest-based company is an innovative corn dry mill technology provider and ethanol plant builder/owner. Their proposal will demonstrate the benefits of integrating an innovative corn waste to ethanol biochemical process into an existing dry corn mill infrastructure.
Participants: E. I. du Pont de Nemours and Company; Novozymes North America, Inc.; National Renewable Energy Laboratory
Production:
- 125 million gallons/year of ethanol, of which roughly 25 percent will be from lignocellulosics.
- Ethanol from lignocellulosic stream and ethanol, chemicals and animal feed from the dry-grind operation.

Technology & Feedstocks
- Integrating production of ethanol into a dry grind corn mill process
- Processing 842 tons/day of corn fiber, corn stover (cobs and stalks)

State of Readiness:
- Broin is building a demonstration plant in order to further develop integrated technologies with DuPont and Novozymes.
- Construction of fully integrated facility starts CY 2007; with a 30 month construction timeline.

Iogen Biorefinery Partners LLC

Corporate HQ: Arlington, VA
Proposed Facility Location: Shelley, Idaho
Description: This project from a leading enzyme player will demonstrate a scaled up biochemical process with the flexibility to process a wide range of agricultural residues into cellulose ethanol.
Participants: Iogen Corporation, Goldman Sachs; Royal Dutch Shell Oil Company; Others
Production:
- 18 million gallons/year in the first plant, 250 million gallons/year in future plants
- Cellulose ethanol & co-products in first plant; future plants to be primarily cellulose ethanol

Technology & Feedstocks:
- Agricultural residues: wheat straw, barley straw, corn stover, switchgrass and rice straw

State of Readiness:
- Tested the overall process at demonstration semi-works plant scale
- Need to test at a commercial scale to validate technologies sufficiently to allow multi-plant rollout at commercial scale
- Construction Starts: 2008; Project completion: end of CY 2010

Range Fuels Inc. (formerly Kergy, Inc.)

Corporate HQ: Broomfield, CO

Proposed Facility Location: Near Soperton, Treutlen County, Georgia
Description: This venture has developed a promising thermo-chemical conversion process, whose success could expand the range of feedstocks available for ethanol production.
Participants: Merrick and Company, PRAJ Industries Ltd., Western Research Institute, Georgia Forestry Commission, Yeomans Wood and Timber; Truetlen County Development Authority; BioConversion Technology; Khosla Ventures; CH2MHill, Gillis Ag and Timber
Production:
- 10 million gallons/year from first unit; ~40 million gallons/year of ethanol and about 9 million gallons/year of methanol from commercial unit
- Ethanol & Methanol

Technology & Feedstocks:
- Conversion of 1200 tons/day of unmerchanteable timber and forest residues
- Catalytic upgrading of syngas to ethanol and methanol

State of Readiness:
- Demonstrated a partially integrated process at 5 tons/day in Colorado
- Publicly announced building and operating commercial-scale facilities in Soperton in 2007 for fully integrated process
- Construction Start: CY 2007; Project Completion: CY 2011

Cellulosic Conversion Technologies

Putting laboratory research to the test on a commercial scale will accelerate advances. Just as with early petroleum refining and corn kernel biorefineries, technologies will be evaluated and refined. Costs will come down and efficiency will rise. It will become increasingly clear which technologies are suited for various feedstocks and regions.

The dominate technologies being developed for cellulosic biorefineries can be classified as biochemical or thermochemical. Some biorefineries will use a combination of methods to maximize efficiency.

Biochemical Methods

With biochemical methods, a pretreatment process liberates sugars from cellulose or hemicellulose. The sugars can then be converted to ethanol or other products using standard methods. Leftover lignin can be burned to provide heat and electricity for the biorefinery. The U.S. DOE Biomass Program is developing a biochemical platform involving the following steps:

Biomass Handling: Biomass goes through a size-reduction step to make it easier to handle and to make the ethanol production process more efficient. For example, agricultural residues go through a grinding process and wood goes through a chipping process to achieve a uniform particle size.

Biomass Pretreatment: In this step, the hemicellulose fraction of the biomass is broken down into simple sugars. A chemical reaction called hydrolysis occurs when dilute sulfuric acid is mixed with the biomass feedstock. In this hydrolysis reaction, the complex chains of sugars that make up the hemicellulose are broken, releasing simple sugars. The complex hemicellulose sugars are converted to a mix of soluble five-carbon sugars, xylose and arabinose, and soluble six-carbon sugars, mannose and galactose. A small portion of the cellulose is also converted to glucose in this step.

Enzyme Production: The cellulase enzymes that are used to hydrolyze the cellulose fraction of the biomass are grown in this step. Alternatively the enzymes might be purchased from commercial enzyme companies.

Cellulose Hydrolysis: In this step, the remaining cellulose is hydrolyzed to glucose. In this enzymatic hydrolysis reaction, cellulase enzymes are used to break the chains of sugars that make up the cellulose, releasing glucose. Cellulose hydrolysis is also called cellulose saccharification because it produces sugars.

Glucose Fermentation: The glucose is converted to ethanol, through a process called fermentation. Fermentation is a series of chemical reactions that convert sugars to ethanol. The fermentation reaction is caused by yeast or bacteria, which feed on the sugars. As the sugars are consumed, ethanol and carbon dioxide are produced.

Pentose Fermentation: The hemicellulose fraction of biomass is rich in five-carbon sugars, which are also called pentoses. Xylose is the most prevalent pentose released by the hemicellulose hydrolysis reaction. In this step, xylose is fermented using Zymomonas mobilis or other genetically engineered bacteria.

Ethanol Recovery: The fermentation product from the glucose and pentose fermentation is called ethanol broth. In this step the ethanol is separated from the other components in the broth. A final dehydration step removes any remaining water from the ethanol.

Lignin Utilization: Lignin and other byproducts of the biomass-to-ethanol process can be used to produce the electricity required for the ethanol production process. Burning lignin actually creates more energy than needed and selling electricity may help the process economics.[7]

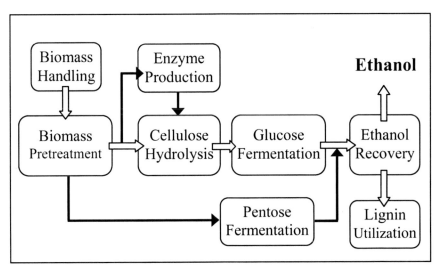

Figure 9-1: The Biochemical Cellulosic Production Process (Adapted from the U.S. DOE Biomass Program)[8]

Thermochemical Methods

Thermochemical methods produce an intermediate synthesis gas (syngas) from biomass by heating with limited oxygen prior to combustion.[9] The syngas can then be converted to ethanol or other end products. Some of the syngas can also be used to produce electrical power needed in the biorefinery. This two-stage process may be effective for more rigid materials such as wood.

A distinct advantage of the syngas fermentation route is its ability to process nearly any biomass resource. Today's corn-based ethanol industry is restricted to processing grain starches. Direct fermentation of biomass, as exemplified by the NREL technology, can handle a wider variety of biomass feedstocks, but more recalcitrant materials lead to high costs. Difficult-to-handle materials, softwoods for example, may best be handled with the syngas fermentation approach.[10] (U.S. DOE)

Biogas as a Transportation Fuel

In Chapter 8, we talked about anaerobic digestion of organic biomass for the production of biogas. This biogas can be used in place of natural gas in biorefineries. Biogas itself can also be used as a transportation fuel, and producing it from cellulosic materials is less difficult compared to making ethanol from cellulosic materials.

The main component of biogas is methane. The same is true for the natural gas piped directly to millions of homes across America. Upgraded biogas can be injected into the natural gas pipeline system and replace natural gas in any application, including transportation.

The 2007 Honda Civic GX NGV car, available at select dealers in California and New York State, runs on compressed natural gas (CNG).[11] This means it should be able to run on upgraded biogas as well. Honda Civic GX car owners can purchase a home refueling appliance that easily hooks into existing natural gas lines. Each home supplied with natural gas is a potential CNG fueling site. According to the U.S. Energy Information Agency, there were over 63 million U.S. residential consumers of natural gas in the U.S. in 2005.[12] This is a tremendous piping and infrastructure head start compared to other alternative fuels. According to NGVAmerica, there are over 1,500 NGV fueling stations in the U.S. in addition to refueling appliances in private homes. Over half of these are available for public use.[13]

Natural gas is currently a fossil-fuel product, but this could change. Microgy Inc., a subsidiary of Environmental Power Corporation, began delivery of upgraded biogas to the existing natural gas pipeline grid in March, 2007. The biogas, branded "renewable natural gas," is produced from anaerobic (without oxygen) digestion of manure and other agricultural wastes at Microgy's Stephensville, Texas facility.[14] Additional similar projects are under development.

A study by the European Union's Joint Research Centre found that compressed biogas used for transportation purposes has a higher well-to-wheel efficiency than ethanol or biodiesel.[15] According to an article on Biopact.com, Europe currently imports about 40% of its natural gas from Russia. A study by the Institute für Energetik und Umwelt, based in Leipzig, and by the Öko-Instituts Darmstadt concluded that Europe could replace all natural gas imports from Russia with biogas by 2020.[16]

Advantages of a biogas transportation system might include:

- Cellulosic biogas production would be less technologically complicated and probably less expensive as compared to cellulosic ethanol.
- A distribution infrastructure for biogas already exists—our natural gas pipeline grid.
- Vehicles are already available in some places that can run on upgraded biogas or natural gas.

In terms of production and distribution technologies, cellulosic biogas is more developed than cellulosic ethanol. The market penetration of natural gas vehicles is limited, however, and automobile makers are not marketing them in most areas of North America. For existing vehicles, EPA-approved natural gas conversions kits are available.[17] Motorists could use fossil-based natural gas until more biogas is available. According to NGVAmerica, the average cost of using natural gas is one-third less than conventional gasoline and there are over 150,000 natural gas fueled vehicles on U.S. roads today.[18]

Waste & Coproduct Feedstocks

Cost-effective processing of cellulosic biomass will require innovation and research. The materials themselves are inexpensive. Many are actually waste products. Much of the cost will come from transporting and storing these bulky, heavy materials. One solution involves using cellulosic feedstocks that are already being transported to a central location for other reasons. Candidates might include:

FOOD PROCESSING WASTE: Food wastes often pose a disposal problem. Wastes from potatoes, fruit, cheese whey, and beverages have free sugars or starch, easily processed into ethanol. Cellulosic technology will enable the use of additional food waste streams. A U.S. DOE abstract estimates a potential for 120 million gallons of ethanol production per year from Florida citrus waste alone.[19]

MUNICIPAL SOLID WASTE: We should be recycling or composting things like paper waste, cardboard, grass clippings, and wood trimmings. Most of us, however, still pay someone to haul this stuff to landfills. Why not haul it to a cellulosic biorefinery instead? BlueFire Ethanol Inc. and ALICO Inc. both propose using waste materials as ethanol feedstocks. BlueFire intends to construct a series of ethanol biorefineries on or near landfills, waste collection sites, and waste separation sites for easy access to biomass feedstocks.[20]

FIBER FROM CORN KERNELS: Hydrolysis methods now used in most corn biorefineries can only convert the softer starch portion of the kernel. Cellulosic technology integrated into existing corn biorefineries will use the hard outer portion of the kernel as well. This will increase the ethanol output per acre of corn without the need to transport additional feedstocks. Dr.

Martha Schlicher, Vice President of Engineering and Operations at Renewable Agricultural Energy Inc. and former director of the National Corn-to-Ethanol Research Center, estimates an increase in ethanol output of 10–12% just from kernel fiber conversion.[21] This would mean about 48 additional gallons of ethanol per acre of corn, assuming a grain yield of 145 bushels. DDG animal feed will still be produced. It will have less fiber and a higher protein content, reducing weight and transportation costs.

The proposed DOE-funded Broin Companies (now Poet LLC) project described in this chapter will convert corn kernel fiber, corn cobs, and stover at an existing Iowa dry-mill corn ethanol biorefinery. A similar process could be used for other grains. The USDA's Agricultural Research Service estimates ethanol production from barley could be increased 10–15% by converting barley hulls to ethanol with cellulosic technology.[22]

BAGASSE FROM SUGAR CANE: Brazilian biorefineries squeeze sugar from cane. Leftover bagasse is burned to produce electricity and heat. Just 20% of the bagasse is needed to power the biorefinery.[23] Excess electricity is now sold onto the grid. With cellulosic conversion technology, excess bagasse could instead be made into even more ethanol. The material is already being hauled to the biorefinery, so no additional transportation infrastructure would be needed. According to Dr. Fernando Reinach, CEO of Alellyx and Director of Votarim New Business Ventures (Brazilian companies involved in ethanol), this would be a more efficient use of bagasse compared to burning for power. He predicts a two-fold increase in ethanol output due to bagasse conversion in the next few years. The leftover lignin could still be burned to produce power and heat needed at the biorefinery.[24]

Agricultural Residue Feedstocks

Ethanol could be made from crop residues such as:

- corn stover (stalks, cobs, and husks)

- wheat straw
- rice straw
- milo (grain sorghum) stubble
- bagasse (sugar cane residue)
- cotton gin trash

A system designed for simultaneous grain and residue harvest could cut down on energy use and cost. If cellulosic ethanol proves successful, there will be an economic incentive to develop specialized harvesting equipment.

Some portion of crop residues should be left on the field to minimize soil erosion and maintain fertility. A comprehensive life cycle analysis coordinated by the U.S. Department of Energy found 40–50% of corn stover can be removed without causing a net release of soil carbon. With "no-till" farming practices, more can be taken. "Preliminary findings," says the DOE, "show that—*when done responsibly*—residue collection can offset petroleum fossil CO_2, reduce our dependence on petroleum and still allow carbon sequestration in soils."[25]

The DOE study projected Iowa alone could produce close to 2.1 billion gallons of pure corn stover (crop residue) ethanol at prices competitive with corn kernel ethanol. Based on their assumption of no-till methods and continuous corn, they found soil erosion would be controlled within tolerable limits established by the USDA and soil organic matter would remain stable over the 90-year time frame studied.[26] No-till farming, already widely practiced, involves sowing seed corn through the previous year's stubble without plowing. Plowing leaves soil susceptible to erosion, especially on sloped fields.

Soil organic matter, another variable measured in the DOE study, is an important component of healthy soils. 85% of the crop residue normally left on the ground, says researchers, rots and releases CO_2 into the air. Only 15% is incorporated into the soil as organic matter. That's why 40–50% of the residue can be harvested sustainably, at least with no-till farming.[27] Essentially, residue harvest at that level recovers carbon that would otherwise be lost into the air. The USDA's Agricultural

Research Service (ARS) is conducting a long-term study to develop guidelines for sustainable crop residue removal. The study began in 2006 and is scheduled to end in 2011.[28]

Cutting-edge farmers such as Richard and Sharon Thompson of Iowa have been harvesting corn stover for animal feed while still maintaining soil quality and high yields.[29] The question yet to be answered is whether it will make economic sense to harvest bulky crop residues from a wide area and transport them to a biorefinery.

Dedicated Cellulosic Energy Crops

Dedicated energy crops might include:

- grasses (including miscanthus, switchgrass, and mixed)
- legumes (including soybeans, sweet clover, and alfalfa)
- trees (such as fast growing poplar or willow)
- sweet sorghum (for cellulose and simple sugars)
- Jerusalem artichoke (for cellulose and simple sugars)
- sugar cane (for cellulose and simple sugars)

Crops intended primarily as biofuel feedstocks can be fine-tuned for the purpose. Many will thrive on land unsuitable for row crops. With proper breeding and agronomic practices, dedicated energy crops can yield much more biomass per acre compared to crop residues. This will allow a shorter transportation distance to biorefineries.

SWITCHGRASS: A native of North America, switchgrass is getting attention as a possible cellulosic ethanol crop. The Oak Ridge National Laboratory (ORNL) has this to say about switchgrass:

> It grows fast, capturing lots of solar energy and turning it into lots of chemical energy—cellulose—that can be liquefied, gasified, or burned directly. It also reaches deep into the soil for water, and uses the water it finds very efficiently.

And because it spent millions of years evolving to thrive in climates and growing conditions spanning much of the nation, switchgrass is remarkably adaptable.

Now, to make switchgrass even more promising, researchers across the country are working to boost switchgrass hardiness and yields, adapt varieties to a wide range of growing conditions, and reduce the need for nitrogen and other chemical fertilizers. By "fingerprinting" the DNA and physiological characteristics of numerous varieties, the researchers are steadily identifying and breeding varieties of switchgrass that show great promise for the future.[30]

In January 2007, Tennessee lawmakers approved spending $72.6 million for what Governor Phil Bredesen called a "transformational effort" to make Tennessee a model for biology-based energy production.[31] The long-term goal is production of 1 billion gallons of ethanol per year from switchgrass. This represents 30% of Tennessee's current gasoline consumption.[32]

MISCANTHUS: A fast-growing grass widely researched in Europe, miscanthus performs better than switchgrass in some locations. Since 2002, the University of Illinois maintained side-by-by-side trials of switchgrass and miscanthus at three locations. As of 2005, biomass yields from miscanthus were more than twice as large as from switchgrass.[33]

PRAIRIE CORDGRASS: In the northern Great Plains, prairie cordgrass (a North American native) might be a good cellulosic ethanol feedstock. Researchers at South Dakota State University (SDSU) have already achieved 10 tons/acre/year from unimproved cordgrass varieties—roughly twice the top yield from switchgrass grown in South Dakota. With the help of a federal grant, SDSU researchers plan to map the genes of prairie cordgrass, with the ultimate goal of improving disease resistance, yield, and chemical composition for cellulosic ethanol production.[34]

BANAGRASS: Different crops and cultural methods will be needed for optimum biomass yields in distinct bioregions like Hawaii. Researchers found that banagrass, a variety of elephant grass, could deliver commercial yields of 18–25 tons/acre/year (dry basis) in Hawaii. This surpasses even sugar cane biomass yields.[35]

POPLAR TREES: Researchers at the Oak Ridge National Laboratory (ORNL) began a breeding program for poplar trees in 1978. Biomass yields of 10 tons/acre/year are readily achievable. With additional research, they expect double that yield. "Our goal is to get 20 tons per acre per year of biomass from trees using less water and nutrients," says Tim Tschaplinski, an ORNL plant physiologist and biochemist. "We want these trees to be able to grow in most regions of the United States, even under drought conditions, and to be harvested in six to seven years."[36] According to researchers at Purdue University, 70 gallons of ethanol can be extracted from a ton of poplar wood. At 10 tons per acre, this amounts to 700 gallons/acre/year of ethanol. Researchers expect to reach 1000 gallons/acre/year by altering the lignin composition in poplar wood, easing conversion to ethanol.[37] If biomass yields could be doubled to 20 tons per acre, ethanol yields from poplar could theoretically reach 2000 gallons/acre/year!

Trees do not need to be harvested every year. If energy prices are down in a given year, harvest could be delayed. In addition, wood is dense, reducing transportation costs. Machinery and techniques for harvest and transportation have already been developed for other uses such as the paper industry.

GIANT SOYBEANS: Thomas Devine and Justin Barone, researchers with the USDA's Agricultural Research Service (ARS), are breeding 7-foot tall soybean plants with large stalks particularly suited for cellulosic ethanol production. "The big obstacle to ethanol production," says Barone, "is finding new enzymes that can break down tough cellulose. We figured, why not just use plants with weaker cellulose?"[38] Barone hopes to develop a test for determining the suitability of plant tissues for ethanol production. As a legume, the soybean generally requires no added nitrogen fertilizer. Bacteria living in legume root nodules capture nitrogen from the air which is then available for plant growth. "It takes natural gas to make nitrogen fertilizer," says Devine. "Eliminating it helps ensure that producing ethanol from soybean cellulose uses less energy than producing it from crops that require nitrogen fertilizer."[39]

Perennial legumes such as alfalfa or sweet clover could also be bred for cellulosic ethanol production, with the added advantage of eliminating the need for yearly planting. Many perennial crops develop deeper roots capable of tapping the huge reservoir of nutrients trapped deep beneath the soil surface.

FEEDSTOCKS WITH CELLULOSE & SIMPLE SUGARS: Some of the sugar crops—sugar cane, sweet sorghum, and Jerusalem artichoke tops—leave behind cellulosic biomass following extraction of simple sugars. These residues can be burned for power and steam, but conversion to ethanol by cellulosic technologies might be a more efficient use.

Brazilian ethanol executive Dr. Fernando Reinach calculates over 6.6 billion gallons per year could be added to Brazil's ethanol output by utilizing bagasse (cellulosic residue from sugar cane). Burning lignin left over from the cellulosic conversion process could provide all the energy needed for the biorefinery. Dr. Reinach believes a future theoretical yield of over 2,300 gallons/acre/year is possible in Brazil based on new hybrid sugar cane varieties, cellulose conversion, and improved technology.[40]

In marginal sugar cane regions like parts of North America, bagasse conversion might provide the extra output needed to make the operation profitable. A similar process could be used for cellulosic residues from sweet sorghum or Jerusalem artichoke tops, making them more viable as energy crops across much of North America.

A Sticky Coproduct

High-value coproducts might increase the profitability of cellulosic ethanol production. USDA ARS researcher Paul Weimer believes wood glue could be among them thanks to a single heat-loving microbe, *Clostridium thermocellum*. The nontoxic plant-based adhesive would not only provide additional revenue for biorefineries, but could replace some of the petroleum-based phenol-formaldehyde glues used to manufacture pressed-wood products.[41] Weimer's research could also streamline ethanol production through consolidated bioprocessing, using a single microbe and a single reactor.[42]

Harvest and Transportation of Feedstocks

With most cellulosic ethanol technologies, economies of scale dictate a large biorefinery. However, most cellulosic biomass is bulky and low in energy density, limiting transportation distance and therefore biorefinery size. Wood feedstocks enjoy advantages in this regard. They are relatively dense, taking up less space in storage and transportation. Also, we already know how to efficiently harvest and transport wood thanks to industries such as paper production. Existing harvest machinery and transportation systems could be adapted for ethanol production.

Corn Stover and grasses such as switchgrass and miscanthus lack the density of wood. They take up more space in transportation and storage. Grasses are usually cut, windrowed, and baled. For ethanol production, researchers are developing harvest systems that skip baling, instead chopping grass as it is harvested. Dr. David Bransby of Auburn University proposes chopping grasses into half-inch pieces suitable for biorefineries. He envisions cutting and windrowing switchgrass. A silage harvester would pick up and chop the dry grass, blowing it into a forage wagon. According to Bransby, chopped switchgrass can be stored in piles on the farm for up to 12 months without a loss in quality.[43]

The Pyrolysis Route to Cellulosic Ethanol

The USDA ARS is testing an on-farm system that would heat biomass in the absence of oxygen to produce a "liquid intermediate" known as "pyrolysis oil" or "bio-oil."[44] Tests show bio-oil has about the same energy content as the parent switch-

156   SUSTAINABLE ETHANOL

grass, but with greater density and 60% less weight. This will reduce the cost of shipping to a central biorefinery for further upgrading to ethanol, other transportation fuels, heating fuels, or chemicals. A pyrolysis by-product known as char could power the process. "The results," says the USDA, "show that char yielded would suffice in providing all the energy required for the endothermic pyrolysis reaction process."[45]

Canadian-based Dynamotive Energy Systems Corporation is already producing bio-oil from cellulosic biomass on a commercial scale. Their Guelph, Ontario fast pyrolysis facility is designed to produce 37,000 gallons of biofuel per day (12.2 million gallons per year). "The attractive economics of BioOil," said Andrew Kingston, president and CEO of Dynamotive, "partly derive from the simplicity of the process, heat transformation of biomass into a liquid and char and the fact that residual cellulosic biomass can be processed at smaller, distributed plants that are significantly less costly to build and operate than other biofuel production facilities that require large scale operations to be economical."[46]

Initially, Dynamotive plans to market BioOil for industrial applications such as "industrial power, heating, paper manufacturing and aluminum smelting."[47] The Intermediate BioOil produced at Guelph will be suitable for applications now served by #2 or #6 heating oil. In a second stage of development, Dynamotive plans to process BioOil into ethanol and synthetic diesel.

Pyrolysis may soon enter the U.S. ethanol scene as well. In June 2007, BioCentric Energy and Core Ventures (BCEI/Core) announced a Joint Venture Agreement with Sustainable Power Corporation (SSTP) to develop the Rivera process of Hydrolysis/Pyrolysis for cellulosic ethanol production.[48]

Regional Biomass Processing Centers

Michigan State University's Dr. Bruce Dale proposes a concept that would lower biomass transportation costs by completing densification and pretreatment at regional biomass processing centers positioned close to biomass sources. A system such as MSU's Ammonia Fiber Expansion (AFEX) or the pyrolysis described above would begin the refining process at the processing centers. The pretreated, densified biomass would then be shipped to large biorefineries for final upgrading.[49]

Regional biomass processing centers could be built on a scale that facilitates local ownership, maximizing benefits to farmers and rural economies. It would be a form of value-added processing. Biomass could be processed for sale to electrical power plants and animal feeding operations in addition to biorefineries. According to Dr. Dale, material from regional biomass processing could also be made into automotive parts, composites, furniture, nutraceuticals, proteins, and enzymes.[50]

Pipeline Transportation of Corn Stover Silage

Tom Schechinger, speaking as chairman of the Board of Managers of BioMass Agri Products in 2000, envisioned a series of collection centers surrounding a central biorefinery, similar to the regional biomass processing center model described above.[51] In this case, however, corn stover would be ensiled for storage at the collection centers. When needed, silage would be washed from the bunker into a pipeline and pumped directly to the biorefinery. Schechinger noted several potential benefits of biomass pipelines:

- makes larger biorefineries feasible
- Stover does not need to be dried prior to storage.
- Pipelines bypass weather and truck congestion problems.
- Stover slurry pipelines would be much less expensive to build as compared to petroleum pipelines. They would be more akin to standard water lines.

Researchers at the University of Alberta have studied the possibility of beginning saccharification in a pipeline on the way to the biorefinery. In the case of larger biorefineries, they predict lower costs as compared to a mid range of truck transportation costs.[52] These researchers also see promise in transporting wood chips by pipeline.[53]

Economics of Cellulosic Ethanol

We will not know how competitive cellulosic biorefineries can be until they have developed a track record. Even in 2003, though, the DOE seemed confident cellulosic ethanol would eventually be competitive:

> Expected yields from a grassroots biomass syngas-to-ethanol facility with no external fuel source provided to the gasifier, are 70–105 gallons of ethanol per ton of dry biomass fed. The economics of this route appear to be competitive with today's corn-based ethanol and projections for direct fermentation of biomass. One report states projected cash costs on the order of $0.70 per gallon, with feedstock available at $25 per ton.
>
> Capital costs are projected at about $3.00 per gallon of annual capacity. The rational price, defined as the ethanol sales price required for a zero net present value of a project with 100% equity financing and 10% real after-tax discounting, is projected to be $1.33 per gallon. These economics would support a successful commercial project at the current ethanol sales price of $1.00–$1.50 per gallon.[54]

How Much Ethanol Can We Make?

Any displacement of petroleum is welcome, but how much cellulosic ethanol or butanol can be made without unacceptable consequences? Researchers have made estimates. A U.S. DOE study discussed earlier in this chapter found Iowa alone could sustainably produce close to 2.1 billion gallons of ethanol per year from corn stover at prices competitive with the usual corn kernel-derived ethanol. This would represent more than half of the total U.S. production of ethanol in 2005.[55]

The U.S. Departments of Agriculture and Energy co-sponsored a major 2005 study commonly referred to as the "Billion Ton Study." The 60-page report was prepared by Oak

Ridge National Laboratory.[56] They estimate 1.3 billion tons of biomass could be made available annually for biofuel production by the middle of the 21st century without damaging our soils or food supply. This would be enough biomass, says the report, to displace more than 30% of current U.S. Petroleum consumption. 998 million tons of biomass would come from agricultural lands and 368 million tons from forestlands. The forestland number is based on assumptions including:

- All forestlands not currently accessible by road were excluded.
- All environmentally sensitive areas were excluded.

The 998 million tons of biomass per year from agricultural resources is based on assumptions including:

- yields of corn, wheat, and other small grains increase by 50%
- the residue to grain ration for soybeans increased to 2:1
- harvest technology capable of removing 75% of annual crop residues (when removal is sustainable)
- all cropland managed by no-till methods
- 55 million acres of cropland, idle cropland, and cropland pastures dedicated to the production of perennial bioenergy crops
- all manure in excess of that which can be applied on-farm for soil improvement under anticipated EPA restrictions utilized for biofuel
- all other available residues utilized

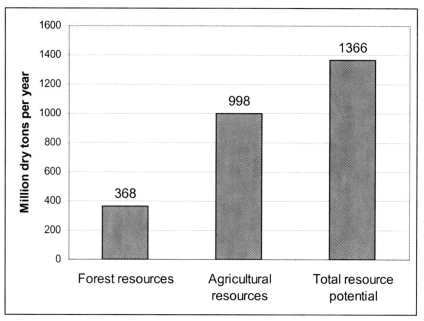

Figure 9-2: Annual Biomass Potential from U.S. Forests and Agriculture (Source: ORNL "Billion Ton Study"[57])

The "billion ton study" is based on relatively modest changes in land use. The authors of the study believe more advanced production techniques could push sustainable biomass availability even higher. The commercialization of cellulosic bioprocessing is just beginning, but the future looks promising.

Notes

1. U.S. DOE Energy Efficiency and Renewable Energy Biomass Program, *Information Resources*, 2007, http://www1.eere.energy.gov/biomass/abcs_biofuels.html.
2. Ibid.
3. U.S. Army Natick Development Center, "Enzymatic Conversion of Waste Cellulose," as reproduced in Michael Wells Mandeville, *Solar Alcohol: The Fuel Revolution* (Ambix: Port Ludlow, WA, 1979), 41.
4. USDA Agricultural Research Service, "Breaking Down Walls," *Agricultural Research Magazine*, April 2007, http://www.ars.usda.gov/is/AR/archive/apr07/walls0407.htm

5. U.S. Department of Energy, "DOE Selects Six Cellulosic Ethanol Plants for Up to $385 Million in Federal Funding," news release, February 28, 2007, http://www.energy.gov/news/4827.htm.
6. Ibid.
7. U.S. DOE Energy Efficiency and Renewable Energy Biomass Program, *Information Resources*
8. Ibid.
9. U.S. DOE Energy Efficiency and Renewable Energy Biomass Program, *Technologies*, 2005, http://www1.eere.energy.gov/biomass/thermochemical_platform.html.
10. P.L. Spath and D.C. Dayton, "Preliminary Screening —Technical and Economic Assessment of Synthesis Gas to Fuels and Chemicals with Emphasis on the Potential for Biomass-Derived Syngas," *National Renewable Energy Laboratory,* 2003, http://www.nrel.gov/docs/fy04osti/34929.pdf.
11. American Honda Motor Co., Inc., *2007 Civic GX NGV,* 2007, http://automobiles.honda.com/models/model_overview.asp?ModelName=Civic+GX.
12. U.S. Energy Information Administration, *Number of Natural Gas Consumers,* http://tonto.eia.doe.gov/dnav/ng/ng_cons_num_dcu_nus_a.htm.
13. Natural Gas Vehicles for America, http://www.ngvc.org/about_ngv/index.html
14. Environmental Power Corporation, "Environmental Power Announces First Delivery of Pipeline Natural Gas From Huckabay Ridge Facility," news release, March 26, 2007, http://ir.environmentalpower.com/releaseDetail.cfm?ReleaseID=235381.
15. Biopact, *Hydrogen Out, Compressed Biogas in*, October 1, 2006, http://biopact.com/2006/10/hydrogen-out-compressed-biogas-in_01.html.
16. Biopact, *Study: biogas can replace all EU imports of Russian gas by 2020,* February 10, 2007, http://biopact.com/2007/02/study-biogas-can-replace-all-eu-imports.html.
17. U.S. DOE Alternative Fuels Data Center, *Conversion Company Industry Contacts*, April 17, 2007, http://www.eere.energy.gov/afdc/progs/res_guide.cgi?CONVCO
18. Natural Gas Vehicles for America, http://www.ngvc.org/about_ngv/index.html.
19. U.S. Department of Energy Office of Energy Efficiency and Renewable Energy, *Citrus Waste Biomass—Inventions & Innovation Project Abstract*, http://www.eere.energy.gov/inventions/pdfs/renewablespirits.pdf.
20. BlueFire Ethanol, "BlueFire Ethanol Fuels Co-Sponsors the 4th Annual California Biomass Collaborative Forum March 27–29," news release, March 27, 2007, http://www.marketwire.com/mw/release_html_b1?release_id=231395.
21. Martha Schlicher, "Biofuels in the U.S.: Today and in the Future," Presentation at the *4th annual Life Sciences & Society Symposium* at the University of Missouri, Columbia, March 15, 2007.

22. USDA Agricultural Research Service, *Economic Competitiveness of Renewable Fuels Derived from Grains and Related Biomass 2006 Annual Report*, 2007.
23. Fernando Reinach, "Biofuels in Brazil: Today and in the Future," Presentation at the *4th annual Life Sciences & Society Symposium* at the University of Missouri, Columbia, March 15, 2007.
24. Ibid.
25. U.S. DOE National Renewable Energy Laboratory, *Life-Cycle Analysis of Ethanol from Corn Stover*, March 2002, http://www.nrel.gov/docs/gen/fy02/31792.pdf.
26. John Sheehan et al, "Energy and Environmental Aspects of Using Corn Stover for Fuel Ethanol," *Journal of Industrial Ecology* 7 (Summer/Fall 2003): 117–146.
27. U.S. DOE National Renewable Energy Laboratory, *Life-Cycle Analysis of Ethanol from Corn Stover,* March 2002, http://www.nrel.gov/docs/gen/fy02/31792.pdf.
28. USDA Agricultural Research Service, *Impact of Residue Removal for Biofuel Production on Soil— Renewable Energy Assessment Project*, 2007, http://www.ars.usda.gov/research/projects/projects.htm?accn_no=410653.
29. Richard, Sharon, Rex, and Lisa Thompson, *Alternatives in Agriculture: Thompson on Farm Research*, 2004, http://www.pfi.iastate.edu/ofr/Thompson_OFR/TOC_Thompson.htm
30. U.S. DOE Oak Ridge National Laboratory, *Biofuels from Switchgrass: Greener Energy Pastures*, http://bioenergy.ornl.gov/papers/misc/switgrs.html.
31. DOE Oak Ridge National Laboratory's Communications and External Relations, "Tennessee Steps Up," *Oak Ridge National Laboratory Review,* 2007, http://www.ornl.gov/info/ornlreview/v40_1_07/article09.shtml.
32. Ibid.
33. Emily A. Heaton, "Miscanthus Bioenergy: Achieving the 2015 Yield Goal of Switchgrass," *Miscanthus at the University of Illinois*, http://miscanthus.uiuc.edu/.
34. South Dakota State University, "Prairie Cordgrass for Cellulosic Ethanol Production," Newswise news release, June 8, 2007, http://www.newswise.com/articles/view/530724/.
35. Charles M. Kinoshita and Jiachun Zhou, "Siting Evaluation for Biomass-Ethanol Production in Hawaii," Prepared for the National Renewable Energy Laboratory by the *Department of Biosystems Engineering College of Tropical Agriculture and Human Resources University of Hawaii at Manoa*, October 1999, http://www.hawaii.gov/dbedt/ert/new-fuel/files/bioethanol/ch03.html.
36. Carolyn Krause, "The People's Tree," *Oak Ridge National Laboratory Review*, http://www.ornl.gov/info/ornlreview/v40_1_07/article04.shtml.
37. Susan A. Steeves, "Fast-growing Trees Could Take Root as Future Energy Source," Purdue University news release, August 23, 2006, http://www.purdue.edu/.

38. Don Comis, "Turning Soybean Plants into Ethanol or Particleboard," *Agricultural Research Magazine*, November/December 2006, http://www.ars.usda.gov/is/AR/archive/nov06/soybean1106.htm.
39. Ibid.
40. Fernando Reinach, *"Biofuels in Brazil: Today and in the Future."*
41. Ibid.
42. Erin Peabody, "Single Microbe Yields Ethanol, Plus Eco-Friendly Glue," *USDA Agricultural Research Service News and Events,* April 12, 2007, http://www.ars.usda.gov/is/AR/archive/apr07/walls0407.htm.
43. Daniel Davidson, "Using Switchgrass in Cellulosic Ethanol," *DTN Ethanol Center,* February 2, 2007, http://www.dtnethanolcenter.com/.
44. USDA Agricultural Research Service, *Interperative Summary of Pilot-scale Fluidized-bed Pyrolysis of Switchgrass for Bio oil Production,* April 3, 2007, http://www.ars.usda.gov/research/publications/Publications.htm?seq_no_115=202978.
45. Ibid.
46. Dynamotive Energy Systems Corporation, "Dynamotive Starts Commissioning Intermediate Biooil Plant in Guelph, Ontario, Aiming at 'untapped' Industrial Fuels Market," news release, March 6, 2007, http://www.dynamotive.com/english/news/releases/2007/march/070306.html.
47. Ibid.
48. BioCentric Energy Inc., "BioCentric Energy and Core Ventures (BCEI/Core) Announce Joint Venture Agreement With Sustainable Power Corp. (SSTP)," Marketwire news release, June 4, 2007, http://www.sustainablepower.com/pr060407.html.
49. Bruce E. Dale, "Biofuels: Thinking Clearly about the Issues," Presented at the 4th *Annual Life Sciences & Society Symposium* at the University of Missouri, Columbia, March 14, 2007; Also see www.everythingbiomass.org.
50. Ibid.
51. Tom Schechinger, "Corn Stover Collection Methods—Present and Future," Presentation at the *Bioenergy 2000 Conference,* 2000, http://devafdc.nrel.gov/pdfs/4922.pdf.
52. Amit Kumar, Jay B. Cameron and Peter C. Flynn, "Pipeline Transport and Simultaneous Saccharification of Corn Stover," *Bioresource Technology* 96 (May 2005): 819–829.
53. Amit Kumar, Jay B. Cameron, and Peter C. Flynn, "Large-scale Ethanol Fermentation through Pipeline Delivery of Biomass," *Applied Biochemistry and Biotechnology,* Spring 2005.
54. P.L. Spath and D.C. Dayton, "Preliminary Screening—Technical and Economic Assessment of Synthesis Gas to Fuels and Chemicals with Emphasis on the Potential for Biomass-Derived Syngas," *U.S. Department of Energy National Renewable Energy Laboratory*, December 2003, http://www.nrel.gov/docs/fy04osti/34929.pdf.
55. Sheehan et al, "Energy and Environmental Aspects of Using Corn Stover"

56. Robert Perlack, Lynn Wright, Anthony Turhollow and Robin Graham (Oak Ridge National Laboratory), Bryce J. Stokes (Forest Service—U.S. Department of Agriculture), and Donald C. Erbach (Agricultural Research Service—U.S. Department of Agriculture), "*Biomass as Feedstock for a Bioenergy and Bioproducts Industry: The Technical Feasibility of a Billion-Ton Annual Supply,*" Sponsored by *U.S. Departments of Energy and Agriculture,* April 2005.
57. Ibid.

Chapter 10

ENERGY BALANCE: IS ETHANOL RENEWABLE?

In order to convert matter or energy from one form to another or move it from one place to another, energy must be expended. This principle is behind the concept of energy balance or net energy ratio—energy obtained from a system divided by the energy put into a system. It's supposed to help us compare different energy systems. Unfortunately, net energy ratio is not an ideal comparison tool because of the different qualities of various energy inputs and outputs.

All BTU's are Not the Same

When computing net energy ratio, we need to assign some unit of energy to each energy carrier involved. Generally, energy in and out is measured in British Thermal Units (BTU's). One BTU is the heat energy needed to raise the temperature of one pound of water from 60°F to 61°F at one atmosphere pressure.[1] The number of BTU's for an energy carrier such as ethanol or gasoline, then, is a measure of how well it can heat water when burned. This is probably as good a common denominator as we could use, but it does not take some important factors into account. "All BTU's are not created equal," as Dr. Bruce Dale of Michigan State University puts it.[2]

Different energy carriers are not interchangeable. A pile of coal with the same BTU content as a gallon of gasoline will not get you down the road if you put it in your car's gas tank. A BTU worth of gasoline will cost you more than a BTU worth of coal because of our thirst for liquid transportation fuels. If we are going to make comparisons based on net energy ratio, we should do so in a way that reflects our goals—our reasons for wanting to replace gasoline with an alternative like ethanol. Dr. Dale proposes two such "metrics"—Fossil Energy Replacement Ratio and Petroleum Replacement Ratio.[3]

Fossil Energy Replacement Ratio

It is the fossil energy inputs that are non-renewable and can cause pollution of water and air. That's why most energy balance studies actually consider the Fossil Energy Replacement Ratio (FER) rather than the total energy ratio, whether they call it that or not. In order to calculate FER, energy delivered to the consumer is divided by fossil energy inputs.

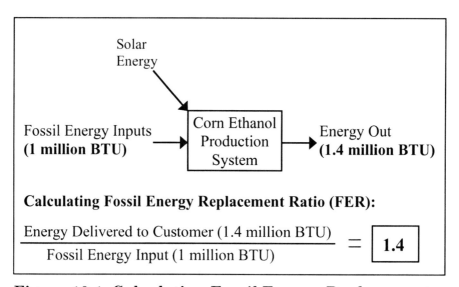

Figure 10-1: Calculating Fossil Energy Replacement Ratio for Corn Ethanol (Concept from Dale, 2007, using data from J. Sheehan & M. Wang, 2003)[4]

The majority of recent studies looking at ethanol from corn kernels show a positive FER (over 1.0), meaning more BTU's are available in the ethanol than in the fossil fuels that went into producing that same ethanol. Of the 13 major studies on the subject completed between 1998 and 2005, 9 showed a positive net energy balance for corn ethanol.[5] In 2001, the USDA calculated an industry average 1.7 net energy ratio for corn ethanol (19 state average).[6] Dr. Dale uses the conservative 1.4 FER figure in his comparisons. This means 1.4 BTU's are delivered to the end consumer for every 1 BTU of fossil fuel input into the corn ethanol production system.

Higher FER means less fossil fuels were consumed for each available BTU. In other words, an energy carrier with a higher FER displaces more fossil fuels. In calculating FER, BTU's from direct solar energy are not counted against ethanol. We can be fairly certain the sun will continue to shine and plants will continue to convert sunlight into biomass from year to year. Input from sunlight is what gives ethanol a positive Fossil Energy Replacement Ratio. Sunlight also sustained the plant life that became fossil fuels, but that took millions of years.

Petroleum Replacement Ratio

Petroleum is often imported from potentially unstable sources and shipped through areas vulnerable to environmental damage. Most fossil fuel inputs for U.S. ethanol production, on the other hand, are sourced from within North America—mainly coal and natural gas. Petroleum Replacement Ratio (PRR) reflects the degree of reliance on petroleum for a given energy carrier. In order to calculate PRR, energy delivered to the customer is divided by petroleum inputs. It counts only the petroleum BTU input against the BTU's delivered to the final customer. An energy carrier with a higher PRR displaces more petroleum. Dr. Dale calculates a PRR of 20 for ethanol from corn kernels. For every 20 BTU's delivered to the consumer, only 1 petroleum BTU went into the corn ethanol production system. This number reflects the fact that corn ethanol production relies on petroleum to a very small extent, making it desirable for our economy and security.

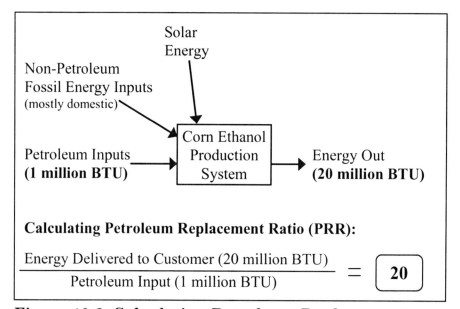

Figure 10-2: Calculating Petroleum Replacement Ratio for Corn Ethanol (Concept from Dale, 2007, using data from Farrell, et al, 2006) [7]

Rating Cellulosic Ethanol

Most experts believe cellulosic ethanol will boast a better fossil energy replacement ratio as compared to corn kernel ethanol based on factors such as:

- Perennial feedstock crops will require fewer fertilizers, pesticides, and herbicides.
- Cellulosic biorefineries can make use of municipal solid waste and other waste streams.
- Un-fermentable lignin can be used as a process fuel in place of natural gas.
- Enough lignin may be left over to also produce electricity.

For his comparisons, Dr. Dale cites a 5.3 Fossil Energy Replacement Ratio for cellulosic ethanol as calculated by J. Sheehan & M. Wang, 2003. Dale calculates a Petroleum Replacement

Ratio of 12.5 for cellulosic ethanol using data from Farrel, et al, 2006.[8] These ratios will improve as cellulosic production methods are perfected.

Comparing Ethanol and Gasoline

Evaluation of an alternative fuel must include comparison to the fuel it replaces. We need to compare the energy balance ratios of ethanol with those of gasoline—our primary transportation fuel. It's easy to overlook this step. Gasoline is a fossil fuel, so our first inclination might suggest it has a fossil energy replacement ratio (FER) of 1. This would ignore the refining process necessary to make crude oil into gasoline. Actually, the FER of gasoline is negative—closer to 0.8. Because most of the inputs are in the form of petroleum, the Petroleum Replacement Ratio of gasoline is also close to 0.8.[9]

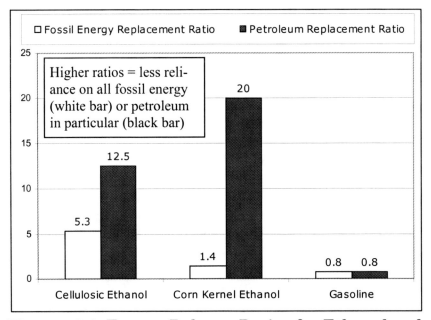

Figure 10-3: Energy Balance Ratios for Ethanol and Gasoline (Adapted from Dale, 2007, using data from Farrell, et al, 2006 and J. Sheehan & M. Wang, 2003) [10]

Figure 10-3 puts together Dr. Dale's numbers for FER and PRR of corn kernel ethanol, cellulosic ethanol, and gasoline. Higher bars (higher ratio numbers) indicate less reliance on the inputs being measured (fossil energy for FER or petroleum energy for PRR). The primary goal of using an alternative fuel is to reduce our reliance on fossil fuels. With significantly higher energy balance ratios compared to gasoline, both cellulosic and corn kernel ethanol accomplish this goal rather well.

Fuel Economy and Energy Balance

For net energy balance, how we use ethanol makes a huge difference. In other words, improvements in fuel economy when using ethanol will improve the energy balance ratios of ethanol.

The BTU content of a transportation fuel is not the only factor determining fuel economy. We are interested in how much work a fuel can accomplish—how far down the road it gets us when we put it in a vehicle. Only a fraction of the energy available in a fuel actually gets transferred to the wheels. The rest is wasted in the form of heat and friction. With the right technology, a greater percentage of ethanol's energy potential will actually be put to work turning a vehicle's wheels. This is mostly because of ethanol's ability to limit engine knock, allowing a higher, more efficient compression ratio. This ability is not reflected in ethanol's BTU content.

In the case of E10 (10% ethanol fuel), very little fuel economy testing has been done, but results of studies by the EPA and the American Coalition for Ethanol indicate some cars actually get *better* fuel economy running on E10 as compared to ethanol-free gasoline.[11] Since most ethanol is currently used in E10, this means ethanol may be displacing fossil fuels more effectively than previously assumed.

In the case of U.S. flex fuel vehicles (FFVs) running on E85 (85% ethanol), the drop in mileage does approximately correspond to ethanol's lower BTU content compared to gasoline (U.S. EPA data). This is because FFVs currently available in the U.S. have not been optimized for E85. Optimization technology already exists. Ford Motor Company, for instance, is considering an ethanol injection concept developed at MIT.[12] Researchers expect ethanol injection to improve fuel economy by 31% over gasoline alone while adding less cost to a vehicle than the premium charged for hybrid electric technology.[13]

If properly accounted for, fuel economy optimization would raise energy balance ratios since less ethanol would be required to replace a given amount of gasoline. When calculating energy balance ratios, more BTU's would need to be credited to each gallon of ethanol. This would more accurately reflect the work ethanol can do in an optimized vehicle in terms of getting on down the road.

A switch to using hydrated ethanol would also improve energy balance ratios. According to the Process Design Center, production of hydrated ethanol requires 45% less energy compared to anhydrous ethanol production.[14] The reduction in energy use would boost ethanol's energy balance ratios. Nearly all fuel ethanol sold in North America is anhydrous so it can be mixed with gasoline for an E10 blend. Almost half of Brazil's fuel ethanol is used in the hydrated form.[15] With the right technology, we could use hydrated ethanol in North America as well (see Chapter 6).

Variables and Trends

Calculating a net energy ratio is not an exact science. Judgments must be made about counting inputs and outputs, and how much energy value will be assigned to each one. How much fossil fuel energy does it take to grow a bushel of corn? Will you consider all corn acres in North America, or just the Corn Belt region? Will you consider the energy it takes to manufacture farm machinery or feed farm workers? Will you go by the average amount of fossil fuel energy required to grow corn and run a biorefinery or the industry best? If you go by averages, your figures might be obsolete in short order because of growing efficiencies in farming and biorefining. Finally, how much energy credit will you allow for coproducts such as animal feed left over from ethanol production? These are just some of the variables involved, illustrating the need to take energy balance figures as "snapshots" based on a particular accounting of inputs and outputs.

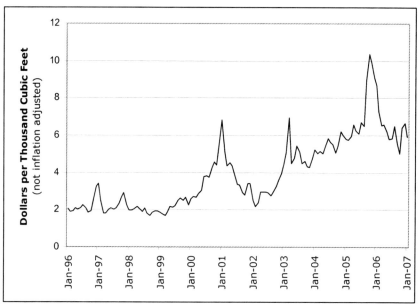

Figure 10-4: U.S. Natural Gas Wellhead Price, 1996–2007 (Data from U.S. Energy Information Administration)

What input and output numbers should we use when it comes to net energy balance? Those of us looking toward the future might lean toward more optimistic numbers because of the long-standing trend toward less energy use in agriculture.[16] Energy use per unit of agricultural output peaked during the 1970's, declining considerably since 1980.[17] This trend will likely continue thanks to higher energy prices. Natural gas prices are trending higher, for instance (figure 10-4). Nitrogen fertilizer is made from natural gas. With crop rotations, precision farming techniques, and alternative cropping systems, many farmers are achieving top yields using less fertilizer and lower inputs in general (see Chapter 7).

In 2006, The National Resource Defense Council and Climate Solutions commissioned the Institute for Lifecycle Environmental Assessment to conduct a research review on the topic of ethanol.[18] They examined 6 studies published since 1990 on corn kernel ethanol and 4 on cellulosic ethanol. Results are summarized in the title, *Ethanol: Energy Well Spent*. They found corn kernel ethanol reduces greenhouse gas emissions and fossil fuel use. The study foresees even better outcomes from cellulosic ethanol, while acknowledging the role of corn kernel ethanol as a foundation for a cellulosic biofuel industry.[19] In North America, corn kernels play a necessary role in biofuel development because they are an abundant existing source of easily fermentable starches. Crops and production systems designed specifically for biofuels will allow dramatic advances in net energy balance and sustainability.

Notes

1. U.S. DOE Energy Efficiency & Renewable Energy, *Biomass Energy Data Book Glossary*, http://cta.ornl.gov/bedb/glossary.shtml.
2. Bruce E. Dale, "Biofuels: Thinking Clearly about the Issues," Presented at the *4th Annual Life Sciences & Society Symposium* at the University of Missouri, Columbia, March 14, 2007; Also see www.everythingbiomass.org.
3. Ibid.
4. Bruce E. Dale, "Biofuels: Thinking Clearly about the Issues"
5. Michael Wang, "The Debate on Energy and Greenhouse Gas Emissions Impacts of Fuel Ethanol," *University of Chicago Argonne National Laboratory*, 2005, 24.
6. Hosein Shapouri, "Net Energy Balance of Biofuels," Presented at the *4th Annual Life Sciences & Society Symposium* at the University of Missouri, Columbia, March 15, 2007.
7. Ibid.
8. Ibid.
9. Ibid.
10. Ibid.
11. KT Knapp, FD Stump, & SB Tejada, "The Effect of Fuel on the Emissions of Vehicles over a Wide Range of Temperatures," *Journal of the Air & Waste Management Association* 43 (July 1998); and American Coalition for Ethanol, *Fuel Economy Study* (2005), http://www.ethanol.org/documents/ACEFuelEconomyStudy.pdf.

12. Nancy Stauffer, "MIT's Pint-sized Car Engine Promises High Efficiency, Low Cost," *Massachusetts Institute of Technology News Office*, October 25, 2006, http://web.mit.edu/newsoffice/2006/engine.html.
13. D.R. Cohn, L. Bromberg, and J.B. Heywood, "Direct Injection Ethanol Boosted Gasoline Engines: Biofuel Leveraging For Cost Effective Reduction of Oil Dependence and CO2Emissions," *Massachusetts Institute of Technology*, April 20, 2005.
14. Process Design Center, "Ethanol in Gasoline –Hydrous E15," presented at the *Conference Sustainable Mobility*, November 21, 2006, http://www.energyvalley.nl/uploads/Mr_Keuken.pdf.
15. Sergio Barros, "Brazil Sugar Annual 2006," *USDA Foreign Agricultural Service GAIN Report Number BR6002,* April 10, 2006.
16. Hosein Shapouri, "Net Energy Balance of Biofuels."
17. Randy Schnepf, "CRS Report to Congress, Energy Use in Agriculture: Background and Issues," *Congressional Research Service*, November 19, 2004, 7–8.
18. Natural Resources Defense Council and Climate Solutions, *Ethanol: Energy Well Spent,* February 2006. http://www.nrdc.org/air/transportation/ethanol/ethanol.asp
19. Ibid.

Chapter 11

FACING OUR ENERGY FUTURE

Alcohol was used as a fuel well before the petroleum era. Over time, it came to be known as power alcohol, agricrude alcohol, gasohol, and finally ethanol. As the era of cheap oil comes to a close, ethanol and other biofuels will play an important role in our energy future. Just as William J. Hale's 1930's "agricrude"[1] terminology sounds odd to us, today's biofuel technology will seem old-fashioned to future generations. We will find new ways to make, distribute, and use biofuels. Based on reasonably achievable advances, the future might include:

- sustainable farming of perennial energy crops
- additional biofuels such as biobutanol (a 4-carbon alcohol), bio-oil, and biogas joining ethanol and biodiesel
- transporting biofuels by pipeline
- super-efficient automobiles getting better fuel economy on biofuels than on petroleum
- "sugar cars" fueled by on-board conversion of sugars and starches to hydrogen[2]
- replacing U.S. petroleum imports with domestic biofuels

Ramping up production and infrastructure will take time, while our appetite for liquid fuel continues to grow. The U.S. Department of Energy (DOE) predicts alternative technologies will displace the equivalent of only 4% of projected U.S. annual consumption of petroleum products by 2015.[3] Such small numbers cause some skeptics to dismiss the whole idea of ethanol and other biofuels. This would be a mistake. Biofuels are not perfect, but they are better than the status quo. With technologies like cellulosic ethanol and biobutanol, biofuel market share will continue to grow. The DOE predicts alternative technologies could displace up to 34% of U.S. petroleum consumption in the 2025 through 2030 time frame, "if the challenges are met."[4]

Balancing our Energy Budget

A sustainable energy future is dependent on the investments, sacrifices, and choices we make as a society and as individuals. It will require both energy conservation and renewable alternatives. It will do little good to produce more biofuels if we simply increase our energy consumption to match. Unfortunately, Americans have demonstrated little self-control when it comes to energy use.

By all means, we should encourage technological advances in biofuel production. Yet production is only half of the biofuel equation. It also matters how efficiently we use those fuels. One of the most overlooked opportunities with ethanol is its largely untapped potential to improve fuel economy. Automakers can implement fuel economy technologies that take advantage of ethanol's knock-suppression properties (see Chapter 6). These technologies can dramatically improve fuel economy, even for the larger vehicles much sought after by the North American motoring public. We need to increasingly emphasize fuel economy if we expect to kick the petroleum habit.

Pressure to use less petroleum and more renewable fuels could come from three main sources:

1. Higher petroleum prices (market forces)
2. Customer demand (our purchasing choices)
3. Government regulations and incentives (our political choices)

Realistically speaking, it all begins with petroleum prices. We might say we're in favor of renewable fuels, but cost is a powerful psychological and practical force. It probably drives our energy purchasing decisions more than we care to admit. Petroleum prices cannot be counted on to remain high enough to induce good far-sighted decisions. Gasoline use will often cost less than using ethanol or other renewable alternatives, forcing us to choose between our principles and our pocketbooks.

In the 1980's and 1990's, it seems we quickly forgot about the 1970's fuel shortages and gas lines. Giving full reign to our addiction, we largely squandered a deluge of cheap oil, with little thought for the future. The U.S. Army, for instance, developed a pre-pilot cellulosic biorefinery in the 1970's. They predicted conversion of cellulosic material to glucose (a sugar) would be practical on a "very large scale" by 1980.[5] Perhaps they were overly optimistic, but how much better off would we now be if we had invested more heavily in cellulosic research decades ago? Have we learned anything as a society? Will we seriously invest in petroleum alternatives? If we want to, we can break our addiction to oil.

Notes

1. William J. Hale, *Farmward March* (New York: Coward-McCann, 1939).
2. Researchers at Virginia Tech, Oak Ridge National Laboratory, and the University of Georgia are developing technology for the direct enzymatic production of hydrogen from sugars and water. If perfected, it could increase the efficiency with which we use biomass such as corn starch. Susan Trulove, "Novel sugar-to-hydrogen technology promises transportation fuel independence," Virginia Tech news release, May 23, 2007, http://www.vtnews.vt.edu/story.php?relyear=2007&itemno=300; The Zhang Lab, http://filebox.vt.edu/users/ypzhang/research.htm.
3. United States Government Accountability Office, *CRUDE OIL: Uncertainty about Future Oil Supply Makes It Important to Develop a Strategy for Addressing a Peak and Decline in Oil Production,* February 2007, http://www.gao.gov/new.items/d07283.pdf.
4. Ibid.
5. U.S. Army Natick Development Center, "Enzymatic Conversion of Waste Cellulose," as reproduced in Michael Wells Mandeville, *Solar Alcohol: The Fuel Revolution* (Ambix: Port Ludlow, WA, 1979), 41.

Ethanol Questions and Answers

- **What is ethanol fuel?**

Ethanol fuel is an alcohol used in internal combustion engines. In North America, it is usually sold blended with gasoline as E10 or E85.

- **What is the difference between E10 and E85?**

E10 is an automotive fuel made up of about 10% ethanol and 90% gasoline. Modern cars can use it without modification. E10 is widely available in North America.

E85 is made up of approximately 85% ethanol and 15% gasoline. It can be used only in vehicles equipped for higher levels of ethanol, such as flex-fuel vehicles. Find your nearest E85 pump by visiting the National Ethanol Vehicle Coalition web site: **www.e85fuel.com**.

A U.S. DOE tool locates all kinds of alternative fuels: **www.eere.energy.gov/afdc/infrastructure/locator.html**.

- **What is a flex-fuel vehicle?**

In North America, "flex-fuel vehicle" (FFV) usually refers to an automobile designed to run on standard gasoline, E85, or any combination of the two. In other words, a flex-fuel vehicle runs fine on anywhere from 0% to 85% ethanol. Many FFVs available in Brazil can run on 100% ethanol (see Chapter 5). The National Ethanol Vehicle Coalition maintains a list of FFVs at **www.e85fuel.com**.

- **Will ethanol hurt my car?**

Most manufacturers specifically approve the use of E10. If your car is very old or antique, you may need ethanol-free gasoline. E85 should be used only in vehicles specifically designed for it, commonly known as flex-fuel vehicles (see Chapter 5).

- **Will ethanol reduce miles per gallon?**

Studies by the EPA and American Coalition for Ethanol indicate E10 reduces fuel economy for some car models and improves it for others. Overall, the results are better than you would expect based on energy content alone.

Using E85 in flex-fuel vehicles (FFVs) currently available in North America causes a significant loss in fuel economy. This is due to ethanol's lower energy density. Future FFV models could get better fuel economy by taking advantage of ethanol's knock-suppression ability (see Chapters 5 and 6). Look up EPA fuel economy data for FFVs at **www.fueleconomy.gov**.

The EPA and DOE provide an online tool for calculating the cost of using E85 and gasoline in your region and using your particular FFV. Click the cost calculator link at **www.fueleconomy.gov/feg/flextech.shtml**.

- **Is ethanol good for the environment?**

Many researchers believe ethanol is better for the environment than the gasoline it replaces. Advances in production and use efficiency are improving that advantage (see Chapter 4).

- **Isn't ethanol really non-renewable because of how much fossil energy it takes to make it?**

A few studies conclude fossil fuel inputs for ethanol production are greater than the energy derived from ethanol use. Most studies, however, show a net gain because of solar energy collected by plants. Thanks to advances in farming and production efficiency, ethanol's net energy balance is getting better (see Chapter 10).

- **Can the U.S. totally replace gasoline with ethanol in the near future?**

Not unless we cut back drastically on our energy use! Ethanol can, however, be a significant factor in reducing our dependence on imported oil. Other renewable solutions like butanol, biodiesel, biogas, direct solar, and wind will also be important. The U.S. Department of Energy predicts "alternative technologies" could displace the equivalent of 4% of projected U.S. annual consumption of petroleum products by 2015, and 34% in the 2025–2030 timeframe.[1]

- **Is ethanol made from anything other than corn?**

Ethanol can be made from any sugar. This may be in the form of simple sugars from sugar cane, sweet sorghum, and sugar beets. Starches from crops like corn, wheat, and milo can be converted to simple sugars for ethanol production as well. Ethanol and other biofuels can also be made from cellulose and hemicellulose broken down into sugars. Cellulose and hemicellulose can be found in organic municipal waste, crop residues, trees, or various energy crops (see Chapters 8 and 9).

Notes

1. United States Government Accountability Office, *CRUDE OIL: Uncertainty about Future Oil Supply Makes It Important to Develop a Strategy for Addressing a Peak and Decline in Oil Production,* February 2007, http://www.gao.gov/new.items/d07283.pdf.

Selected Resources/Bibliography

For an updated list of resources, visit the *Sustainable Ethanol* web site: **www.ethanolbook.com**

Chapter 1 — A Brief History of Ethanol

Books
- *The Forbidden Fuel: Power Alcohol in the Twentieth Century,* Hal Bernton, William Kovarik, and Scott Sklar, 1982.
- *The Prize: The Epic Quest for Oil, Money & Power,* Daniel Yergin, 1993.

Web Site
- **www.runet.edu/~wkovarik/** (Prof. Bill Kovarik's web site)

Chapter 2 — Will Cheap Oil Return?

Book
- *Twilight in the Desert: The Coming Saudi Oil Shock and the World Economy,* Matthew Simmons, 2006.

Report
- *CRUDE OIL: Uncertainty about Future Oil Supply Makes It Important to Develop a Strategy for Addressing a Peak and Decline in Oil Production* (February 2007 Report from the United States Government Accountability Office) www.gao.gov/new.items/d07283.pdf.

Chapter 3 — Economic and Security Benefits

Reports
- *Contributions of the Ethanol Industry to the Economy of the United States,* John M. Urbanchuk, 2007 (Prepared for the Renewable Fuels Association by LECG LLC)
  Find at http://www.ethanolrfa.org
- *Ownership Matters: Three Steps to Ensure a Biofuels Industry that Truly Benefits Rural America,* David Morris, Institute for Local Self-Reliance, 2006. Find at http://www.newrules.org
- *The Hidden Cost of Oil: An Update,* The National Defense Council Foundation, January 8, 2007. Find at http://www.ndcf.org

Web Site
- www.carbohydrateeconomy.org (A project of the Institute for Local Self Reliance)

Chapter 4 — Environmental Impact

Report
- *Does Ethanol Use Result in More Air Pollution?*, David Morris and Jack Brondum, 2000. Institute for Local Self Reliance. Find at www.carbohydrateeconomy.org

Web Site
- http://www.cleanairchoice.org/ (American Lung Association of the Midwest)

Chapter 5 — E10, E85, and Flex-Fuel Vehicles

Web Sites
- www.e85fuel.com (National Ethanol Vehicle Coalition—Includes E85 station locator, listing of flex-fuel vehicles, and guide for determining if your car is flex-fuel)
- www.eere.energy.gov/afdc/infrastructure/locator.html (U.S. DOE station locator—finds multiple alternative fuels)
- www.fueleconomy.gov (U.S. EPA—gives fuel economy ratings for flex-fuel vehicles)
- www.fueleconomy.gov/feg/flextech.shtml (U.S. EPA—Link computes your cost of driving on E85 and standard gasoline)
- http://www.drivingethanol.org/ (Ethanol Promotion and Information Council—Promoting the benefits of ethanol)
- http://www.ethanolmt.org/ (Ethanol Producers & Consumers—non-profit association to support production and use of ethanol)
- http://www.cleanfuelsdc.org/ (Clean Fuels Development Coalition—non-profit organization to promote new technologies and increased production of clean fuels)
- http://www.americanbiofuelscouncil.com/ (American Biofuels Council—coordinates communication & education)

Chapter 7 — Food, Farming and Land Use

Web Sites
- www.leopold.iastate.edu (Leopold Center for Sustainable Agriculture at Iowa State University)

- **www.sare.org** (USDA's Sustainable Agriculture Research and Education)
- **www.attra.org** (National Sustainable Agriculture Information Service)
- **www.newfarm.org** (Rodale Institute)

Chapter 8 — Ethanol Production

Books about Home and Farm-scale Production
- *The Ethanol Fuel Handbook,* Lynn Ellen Doxon, 2001.
- *Forget the Gas Pumps—Make Your Own Fuel,* Jim Wortham, 1979.

Web Sites
- **http://journeytoforever.org/ethanol.html** (Information and links about small-scale, grassroots ethanol production and use)
- **http://www.ethanolrfa.org/** (Renewable Fuels Association—National trade association for the U.S. ethanol industry)
- **http://www.ethanol.org/** (American Coalition for Ethanol—Non-profit association to promote use and production of ethanol)
- **http://www.greenfuels.org/** (The Canadian Renewable Fuels Association—Non-profit organization to promote renewable fuels)

Chapter 9 — Cellulosic Ethanol

Web Sites
- **www1.eere.energy.gov/biomass/abcs_biofuels.html** (U.S. DOE Energy Efficiency and Renewable Energy Biomass Program)
- **http://www.nrel.gov** (National Renewable Energy Laboratory)
- **http://bioenergy.ornl.gov** (Bioenergy Feedstock Information Network)

Chapter 10 — Energy Balance

Report
- *Ethanol: Energy Well Spent,* Literature Review by the National Resource Defense Council, 2006. www.nrdc.org/air/transportation/ethanol/ethanol.asp

Glossary

alcohol—An organic compound with a carbon bound to a hydroxyl group. Examples are methanol, CH_3OH, and ethanol, CH_3CH_2OH.

anhydrous ethanol—Ethanol with almost all water removed.

ARS—USDA, Agricultural Research Service

bagasse—The solid material left after processing sugar cane to produce sugar or ethanol. Often burned for power production, but could be made into more ethanol with cellulosic processing techniques.

biodiesel—A renewable, biodegradable transportation fuel for use in diesel engines. It is produced from organically derived oils or fats. Biodiesel can be used as a component of or replacement for diesel fuel.

bioenergy—Renewable energy derived from organic matter.

biofuel—Liquid, solid, or gaseous fuel produced by conversion of biomass. Examples include woodchips, biodiesel, ethanol from biomass, bio-oil, and biogas.

biogas—A renewable gas derived from decomposing organic material under anaerobic conditions. Normally consists of 50–60% methane. Can be upgraded to replace fossil-based natural gas for automobiles, heating, cooking, generation of electricity, and other uses.

biomass—Organic matter available on a renewable basis. Includes agricultural crops, residues, wood and wood waste, animal wastes, fast-growing trees, municipal waste, and food processing waste.

bio-oil—A liquid fuel produced by the pyrolysis of biomass. Bio-oil can be upgraded to ethanol or other biofuels and materials. As an energy-dense intermediate, it is easier to transport than raw biomass.

biorefineriy—A facility for the production of biofuels.

BTU—British Thermal Unit. A standard unit of measure representing the heat energy required to raise the temperature of 1 pound of water by 1 degree Fahrenheit at sea level.

butanol—($C_4H_{10}O$) An alcohol. Sometimes referred to as "biobutanol" when made from biomass such as grains or cellulosic material. It has a higher volumetric energy content than ethanol and shows promise as an automotive fuel if production costs can be brought down.

cellulose—A long chain of glucose (sugar) molecules. Strengthens the cell walls of most plants.

cellulosic ethanol—Ethanol made from cellulose or hemicellulose after breaking them down into constituent sugars.

CHP—Combined Heat and Power. Also known as cogeneration. The production of electricity and useful thermal energy from a common fuel source, increasing energy efficiency.

coproduct—A product of biomass processing when multiple products are produced from the same feedstock.

crop residue—Organic residue remaining after harvesting a crop.

crop rotation—Alternating the crops grown on a given plot of land. Can improve soil fertility and help reduce pests and diseases.

DDGS—Dried Distillers Grain with Solubles. Nutrient-rich coproduct of ethanol production from grains. Can be a high quality livestock feed.

denatured ethanol—Ethanol made unfit for beverage use.

distillation—Extraction of a volatile component (such as alcohol) by condensation and collection of vapors produced as a mixture is heated.

DOE—United States Department of Energy

E10—A mixture of 10% ethanol and 90% gasoline based on volume.

E85—A mixture of 85% ethanol and 15% gasoline based on volume.

ERS—USDA, Economic Research Service

EIA—DOE, Energy Information Administration

EPA—United States Environmental Protection Agency

ensilage—The process of partially fermenting and storing fresh plant material in an oxygen-poor environment.

enzyme—A protein or protein-based molecule that can speed up chemical reactions in biomass.

ethanol—(CH_3CH_2OH) A colorless, flammable liquid usually produced by fermentation of sugars. Used as a fuel oxygenate and replacement for gasoline. Ethanol is the alcohol found in alcoholic beverages.

ethanol optimization—Designing an engine for better efficiency, performance, or fuel economy when powered by ethanol.

ethanol-livestock integration—Combining livestock and ethanol production in a synergystic fashion, resulting in greater energy efficiency and other benefits. Livestock waste can serve as the process fuel for a biorefinery, for instance, while DDGS is fed to the livestock.

feedstock—Any material converted to another useful form or product. For example, cornstarch can be a feedstock for ethanol production.

FER—Fossil Energy Replacement Ratio. Energy delivered to the final customer divided by the *fossil energy* inputs into the production system for a given energy carrier such as ethanol. FER is a measure of reliance on fossil energy. Higher FER means less fossil energy went into producing that energy carrier.

fermentation—A biochemical reaction that breaks down complex organic molecules into simpler materials. Bacteria or yeasts can ferment sugars to ethanol, for instance.

flex-fuel vehicle—A vehicle that can operate on alternative fuels (usually E85 in North America) or on traditional fuels such as ethanol-free gasoline or a mixture of alternative and traditional fuels.

fossil fuels—Solid, liquid, or gaseous fuels formed in the ground after millions of years by chemical and physical changes in plant and animal residues under high temperature and pressure.

gasification—A chemical or heat process to convert coal, biomass, wastes, or other carbon-containing materials into a gaseous form that can be burned to generate power or processed into chemicals and fuels, including ethanol.

glucose—A simple six-carbon sugar, $C_6H_{12}O_6$. A sweet, colorless sugar that is the most common sugar in nature and the sugar most commonly fermented to ethanol.

greenhouse gases—Those gases, such as water vapor, carbon dioxide, nitrous oxide, and methane, that are transparent to solar radiation but opaque to long wave radiation.

hemicellulose—Short, highly branched chains of sugars. In contrast to cellulose, which is a polymer of only glucose, a hemicellulose is a polymer of five different sugars. Compared to cellulose, the branched nature of hemicellulose renders it relatively easy to hydrolyze.

hydrated ethanol—Ethanol containing a significant amount of water. Can be used as a fuel in engines designed or modified for the purpose.

hydrolysis—The conversion of a complex substance into two or more smaller units, such as the conversion of cellulose into glucose sugar units for cellulosic ethanol production.

landfill gas—A bio-gas produced as a byproduct of decomposition in landfills. Can be a process fuel or feedstock for biofuel production.

legumes—Plants in the pea family, characterized by their ability to host nitrogen-fixing bacteria, lessening or eliminating the need for nitrogen fertilization. Includes beans, alfalfa, and many native prairie plants.

lignin—The major structural constituent of wood and other plant materials. It cements cells together. A co-product of some cellulosic ethanol production systems, it can be burned as a process fuel.

MIT—Massachusetts Institute of Technology

MTBE—Methyl Tertiary Butyl Ether. A colorless, flammable liquid. Used as an oxygenate additive to gasoline to increase octane and reduce engine knock.

ORNL—Oak Ridge National Laboratory

octane rating—A number indicating a fuel's resistance to self-ignition, hence also a measure of the antiknock performance of the fuel.

oxygenates—Substances which, when added to gasoline, increase the amount of oxygen in that gasoline blend. Ethanol, Methyl Tertiary Butyl Ether (MTBE), Ethyl Tertiary Butyl Ether (ETBE), and Methanol are some examples.

particulate matter—A small, discrete mass of solid or liquid matter that remains individually dispersed in gas or liquid emissions.

perennial crops—Crops that live multiple years without replanting.

petroleum—A generic term applied to fossil-derived oil and oil products in all forms. Includes crude oil, unfinished oil, refined petroleum products, and natural gas plant liquids.

proof—A measure of the volume of ethanol in a liquid. Proof is twice the percentage number, so 100% ethanol would be 200 proof.

PRR—Petroleum Replacement Ratio. Energy delivered to the final customer divided by the *petroleum* inputs into the production system for a given energy carrier such as ethanol. PRR is a measure of reliance on petroleum, a fossil energy source more likely to be imported into the U.S. as compared to other fossil fuels. Higher FER means less petroleum went into producing that energy carrier.

starch—A molecule composed of long chains of linked glucose molecules. Because of the way the glucose molecules are linked, starch can be readily broken down into glucose by enzymes. Glucose, a type of sugar, can then be fermented for ethanol production.

sustainable farming—Using farming methods that maintain productivity, soil fertility, and a healthy ecosystem over the long term.

USDA—United States Department of Agriculture

Index

A

Abengoa Bioenergy, 106, 120, 139
acceleration, 52
addiction to oil, 11, 12, 28, 89, 177
Adkins Energy LCC, 122
Agricultural Research Service, 41, 46, 94, 102, 139, 151, 153, 154, 155
agricultural residues, 93, 137, 140-144, 154, 159
agriculture, 77, 83, 84, 87-91, 102, 119, 127, 158, 160, 172
Agrol, 14, 15
air pollution, 44, 45, 46, 51, 75, 88, 121
alcohol, 12-16, 25, 33, 44, 51, 55, 72, 76, 99, 116, 128, 175, 178
Alellyx, 149
alfalfa, 94, 151, 154
algae, 113, 114, 126
ALICO Inc., 140, 148
alternative process fuels, 119
American Coalition for Ethanol, 53, 63, 66, 67, 170, 179
American Lung Association of Minnesota, 61
Ammonia Fiber Expansion, 157
anhydrous ethanol, 76, 77, 78, 127, 171
animal feeds, 85, 112, 115, 116
antifreeze, 53
anti-knock, 13
antique cars, 53
aqueous ammonia fertilizer, 125
arbuscular mycorrhizal fungus, 94
Argonne National Laboratory, 40, 41, 71
Arizona, 114, 126
aromatics, 43
Asahi Breweries, 104
Asia, 24, 26, 27
Atchison, Kansas, 14, 15
Auburn University, 96, 155

B

bacteria, 94, 109, 116, 139, 144, 145, 153
bagasse, 73, 103, 104, 149, 150, 154
baling, 155
banagrass, 152
barley, 101, 102, 115, 116, 142, 149
barley straw, 142
Barone, Justin, 153
batteries, 72, 73, 74, 79
benzene, 43, 44, 116
beverage, 12, 13, 99, 107, 113
billion ton study, 160
Bio Processing Technology, 86
biodegradable plastic, 86
biodiesel, 41, 56, 72, 73, 113, 114, 116, 126, 147, 175, 180
biodiversity, 96
bioenergy, 32, 76, 79, 99, 113, 114, 121, 125, 126, 129, 130, 137, 159, 160, 175, 176, 180
biofuels, 32, 76, 79, 99, 100, 108, 113, 114, 121, 124-126, 128-130, 137, 151, 156, 159, 160, 175, 176, 180
biogas, 107, 119, 122, 125, 126, 146, 147, 175, 180
biomass, 41, 46, 92, 93, 95, 99, 102, 104, 113, 119, 121-123, 126, 137-139, 141, 144-146, 148, 151-160
BioMass Agri Products, 157
biomass handling, 144
biomass pipelines, 157
biomass pretreatment, 144
bio-oil, 156
biorefineries, 15, 17, 31, 37, 73, 78, 85, 86, 101, 103, 105-107, 112-123, 125, 126, 128, 129, 139, 143, 144, 148, 149, 154-158, 177
bitumen, 25, 34
black liquor, 137
blendstock, 44, 53, 130
BlueFire Ethanol Inc., 140, 148
Brann, Dan, 116
Bransby, Dr. David, 155
Brazil, 42, 77, 78, 102, 103, 118, 128, 154, 171, 178
British Petroleum, 27, 28, 63, 129, 130

British Thermal Unit, 59, 66, 69, 91, 122, 129, 165-167, 170, 171
Broin Companies, 112, 141, 149
Brooks C&D Landfill, 120
bubbling bed fluidized gasifier, 126
butanol, 41, 44, 72, 99, 113, 128-130, 137, 158, 175, 176, 180

C

California, 43, 44, 140, 141, 146
Canada, 25, 34, 54, 58, 102, 106, 107, 115, 119, 156
carbon dioxide, 40, 55, 94, 95, 109, 113, 114, 120, 121, 124, 126, 144, 150
carbon monoxide, 18, 44, 55
carbon sequestration, 150
Carnegie Mellon University, 117
cell structure, 138
cellulase, 144
cellulose, 93, 100, 103, 137-139, 142, 144, 151, 153, 154, 160, 177, 180
cellulose hydrolysis, 144
cellulosic conversion technologies, 143
cellulosic ethanol, 37, 41, 46, 73, 92-94, 96, 99, 100, 107, 121, 137-139, 143, 145-156, 158, 160, 168, 170, 173, 176, 177
cellulosic feedstocks, 93, 100, 137, 148
char, 156
cheap oil, 11, 15, 17-19, 23-25, 65, 88, 89, 175, 177
cheese whey, 101, 107, 148
chemurgy, 14
Chevrolet Volt, 73
China, 27
Chippewa Valley Ethanol Company, 119
chokepoints, 35
citrus peel, 140
Civil War, 12
climate change, 40, 120, 124
Climate Solutions, 173
closed loop biorefineries, 124, 125
Clostridium thermocellum, 154
coal, 28, 86, 121, 122, 165, 167
cold weather starting, 51
combined heat and power, 122-124
co-mingling, 44

compressed natural gas, 146
compression ratio, 13, 42, 70, 74, 170
congress, 29
Congressional Research Service, 45
conservation, 17, 65, 96, 176
constant rpm, 73
Consumer Price Index, 84
coproducts, 85, 86, 105, 107, 109, 112, 113, 116, 125, 126, 129, 140, 154, 172
corn, 34, 40, 41, 46, 83-89, 91-93, 96, 99, 100, 101, 105, 106, 108-115, 117, 119, 125, 139, 141-143, 148-151, 155, 157-159, 166-168, 170, 171, 180
corn cobs, 142, 149
corn exports, 86, 87, 92
corn fiber, 86, 142
corn oil, 86, 113
corn planting, 92
corn prices, 84, 85, 92, 93
corn stover, 45, 93, 139, 142, 149, 150, 151, 157, 158
cosolvency, 44
cost calculator, 59, 179
cotton gin trash, 150
cover crops, 89-91
crop residues, 37, 45, 93, 119, 149, 150, 151, 159, 180
crop rotation, 89, 90, 96, 172
crop scouting, 90
crop subsidies, 31
crotalaria, 93
CSEA Co-Operative Inc., 107, 119

D

dehydration, 78, 110, 145
denatured, 51, 55, 58, 66, 78, 110
Devine, Thomas, 153
direct injection, 72, 74, 75, 79
distillation, 99, 106, 108, 109, 116, 117
distribution, 15, 75, 110, 128, 147
Dr. Bruce Dale, 156, 157, 165, 166, 167-168, 170
Dried Distiller's Grain, 87, 113, 119, 149
Dried Distillers Grains with Solubles, 85, 112, 116
driving cost, 59, 60
dry fractionation, 112, 118

dry milling, 85, 109, 112, 113, 122, 141
DuPont, 141, 142
Dynamotive Energy Systems Corporation, 156

E

E10, 44, 51-55, 59, 63, 65-69, 77, 79, 170, 171, 178, 179
E100, 59, 73
E^3 Biofuels LLC, 124
E85, 45, 51, 54-56, 58-63, 65, 68-76, 78, 79, 170, 178, 179
E85 optimization, 54, 71
E95, 45
East Kansas Agri Ethanol, 123
Eastern Europe, 25
easy oil, 25
economy, 14, 17, 28, 31, 32, 35, 36, 40, 41, 53, 54, 59-63, 65-67, 69, 87, 102, 108, 118, 123, 129, 145, 150, 151, 156, 158, 167, 170, 176, 179
Ecosense Solutions, 75
education, 32
E-Flex, 73
eighteenth amendment, 13
electric motors, 74, 79
electricity, 65, 72, 73, 95, 103, 107, 108, 119, 121, 122, 138, 140, 144, 145, 149, 168
endothermic pyrolysis, 156
EnerGenetics International Inc., 86, 105, 129
energy balance, 40, 65, 66, 74, 75, 91, 95, 103, 105, 112, 114, 118, 121, 123-125, 165-167, 169-172, 179
energy balance ratios, 39-41, 65, 66, 74, 75, 91, 103, 105, 112, 121, 123-125, 165-167, 169, 170
energy crops, 93, 94, 137, 151, 154, 180
energy density, 24, 53, 56, 65, 155, 179
energy efficiency, 65, 116
energy farming, 96, 103
energy future, 175
energy recovery system, 117
energycane, 140
engine knock, 13, 42, 53, 170
engines, 12, 13, 24, 40, 42, 51-53, 56, 61, 68-76, 78, 127, 129, 170, 178

enhanced oil recovery, 33, 114
ensilage, 105
environment, 12, 18, 39, 47, 62, 65, 84, 88, 91, 93, 120, 179
enzyme hydrolysis, 109
enzyme production, 144
enzymes, 109, 116, 141, 144, 153, 157
EROEI, 24
ethanol, 11, 15, 17, 18, 32, 33, 39-46, 51-56, 58-63, 65-79, 83-89, 91-93, 95, 96, 99-101, 103-110, 112-129, 137-156, 158, 165-172, 175-180
ethanol blends, 14, 40, 44, 45, 51-56, 58-63, 65-79, 129, 130, 170, 171, 178, 179
ethanol boosting, 74, 75, 118
Ethanol Boosting Systems LLC, 75
ethanol injection, 45, 78, 79, 171
ethanol optimization, 54, 69, 71
ethanol production, 19, 31, 32, 77, 85, 91, 99-104, 107-109, 111-114, 116-119, 121-123, 126-128, 139, 143-145, 148, 149, 152-156, 167, 171, 172, 179, 180
ethanol-livestock integration, 124
ethyl alcohol, 13, 51
Europe, 12, 14, 16, 34, 147, 152

F

farming, 12, 14-18, 31, 34, 41, 46, 78, 83, 84, 86-94, 96, 105, 106, 108, 115, 125, 126, 150, 151, 155, 157, 159, 171
fast pyrolysis, 156
feedstock, 15, 16, 32, 39, 40, 41, 46, 78, 93, 99-101, 105-108, 115-117, 119, 121, 126, 128, 129, 137, 139, 140-144, 146, 148, 149, 151, 152, 154, 155, 158, 168
fermentation, 15, 99, 100, 103, 106, 109, 110, 113, 114, 138, 140, 141, 144, 145, 146, 158, 168
fertilizers, 14, 16, 39, 46, 84, 88-91, 94-96, 104, 115, 120, 125, 140, 152, 153, 168, 172
field peas, 94, 101, 102, 115
Fischer-Tropsch, 95
flex-fuel, 13, 51, 54-58, 61, 62, 68, 71, 72, 74, 75, 77, 78, 129, 178, 179

INDEX

flex-fuel vehicles, 51, 54-58, 61, 62, 68, 71, 75, 77, 78, 178, 179
Florida, 36, 96, 102, 140, 148
Florida State University, 96
fodder beets, 103
food, 63, 83-88, 95, 96, 103, 107, 109, 112, 137, 148, 159
food prices, 84, 85
food wastes, 37, 103, 148
Forbidden Fuel, 11
Ford Motor Company, 56, 70, 75, 171
forestry wastes, 137
Fossil Energy Replacement Ratio, 166-170
fossil fuels, 11-13, 16, 24, 25, 36, 39-42, 44, 47, 54, 74, 76, 84, 88-91, 94, 121, 128, 166, 167, 169-171, 179
fractionation, 109, 112, 113, 126
fruit, 101, 107, 108, 126, 148
fuel additives, 52
fuel alcohol, 11-14, 16, 18, 51
fuel cells, 46, 66, 76, 78, 79, 118
fuel economy, 13, 40-42, 45, 46, 54, 59, 60-62, 65-72, 74, 79, 129, 170, 171, 175, 176, 179
fuel filters, 53
fuel injection, 52, 56
fuel tank, 44, 46, 52, 55, 74, 75

G

gasification, 95, 114
gasohol, 17, 78, 175
gasoline, 12, 13, 15, 17, 18, 32, 36, 39, 40-46, 47, 51-63, 65-75, 77, 78, 92, 110, 127-130, 165, 169, 170, 177, 178, 179
Gehrke, Russel, 76
General Motors, 57, 70-73
glomalin, 94
glucose, 109, 117, 138, 144, 145, 177
gluten, 86
Golden Triangle Energy Cooperative, 118
grain prices, 15, 31, 83, 85-87
grain sorghum, 101, 102, 115, 150
grass, 46, 90, 93-96, 101, 119, 137, 148, 151, 152, 155
grass clippings, 101, 148

Grassroots Energy LLC, 78
gray water, 126
Great Depression, 14, 15
Great Plains, 12, 96, 152
green revolution, 89
greenhouse effect, 40
greenhouse gas, 40, 41, 45, 46, 60, 95, 114, 120, 121
Gross Domestic Product, 31
groundwater, 18, 42, 44
GS CleanTech, 113,
Gulf of Mexico, 36, 95

H

Hale, William J., 14, 15, 175
Harkin, Senator Tom, 128
health, 13, 43, 59, 86
heating oil, 156
Heggenstaller, Andy, 93
hemicellulose, 100, 137, 138, 144, 145
Henry Ford, 12, 14
herbicides, 16, 39, 46, 88-90, 168
high-diversity grassland, 95
history, 11, 42, 103, 116
Honda Civic GX NGV, 146
hurricanes, 36, 37
hybrid vehicles, 66, 72
hydrated ethanol, 74, 76-78, 118, 127, 171
hydrogen, 46, 66, 73, 76, 129, 140, 175
hydrolysis, 99, 109, 117, 138, 144, 145
Hydrous E15, 78

I

Idaho, 140, 142
imported oil, 16, 35, 36
India, 93
industrial wastes, 137
inflation, 31, 33, 83
Institute for Agriculture and Trade Policy, 87, 118
Institute for Lifecycle Environmental Assessment, 173
Institute for Local Self Reliance, 32, 45, 91
International Food Policy Research Institute, 87

investors, 19, 32
Iogen Biorefinery Partners LLC, 142
Iowa, 31, 53, 84, 86, 93, 102, 105, 113, 123, 129, 141, 149-151, 158
Iowa State University, 31, 85, 93, 102, 105
irrigation, 75, 90, 96, 105, 115

J

Japan, 15, 86, 104, 141
Jerusalem artichokes, 99, 101-103, 106, 107, 126, 151, 154
jobs, 13, 31, 32, 114
John M. Urbanchuk, 31
Joint Research Centre, 147

K

Kansas, 14, 58, 114, 120, 122, 123, 127, 139
Kenney, Kevin, 78
Khosla, Vinod, 100
Kingston, Andrew, 156
Kovarik, Bill, 11, 13

L

L.P. Gill Landfill, 120
Lamberty, Ron, 53
lamp fuel, 12
land use, 83, 91, 160
landfill gas, 120, 121
laws, 18, 54, 127, 152
lead, 13, 42, 43, 66, 119, 146
Lee, Bill, 119
legumes, 93-95, 115, 151, 153, 154
Liebig, Mark, 46
life cycle analysis, 45, 150
Lifeline Foods, 112
lignin, 73, 121, 137, 138, 144, 145, 149, 153, 154, 168
lignin utilization, 145
liquid fuels, 25
liquors, 13
lithium-ion batteries, 73
local ownership, 31, 32, 157
Louisiana, 36, 37, 104
Lugar, Senator Richard, 128
Lynd, Lee, 96

M

Maciel, Milton, 42, 102
mandates, 18, 54, 63
McClune, Lee, 105
methane, 55, 120, 122, 126, 146
methanol, 143
Michigan State University, 156
Microgy Inc., 147
Middle East, 26
millet, 107, 119
Minnesota, 54, 58, 63, 95, 118, 119, 123
miscanthus, 151, 152, 155
Missouri, 53, 54, 63, 75, 77, 108, 112, 118, 121, 122, 139
MIT, 72, 74, 75, 171
Model T, 12
molasses, 101, 103
Montana, 54, 63
motor oil, 56
MTBE, 18, 43, 44
multicolumn distillation, 117
municipal solid waste, 37, 137, 168

N

Natick Development Center, 138, 160, 177
National Agricultural Research Centre for Kyushu Okinawa Region, 104
National Ethanol Vehicle Coalition, 57
National Resource Defense Council, 173
natural gas, 16, 18, 25, 26, 29, 36, 43, 76, 86, 94, 99, 107, 116, 119, 120-122, 125, 146, 147, 153, 167, 168, 172
natural gas pipeline grid, 147
Nebraska, 17, 120, 124
net energy balance, 65, 66, 74, 75, 91, 95, 103, 105, 112, 114, 118, 121, 123, 124-126, 165-167, 169-173, 179
NGVAmerica, 146, 148
Nichols, Nancy, 115
nitrogen, 55, 89, 91, 94, 95, 115, 116, 152, 153
North Central Research Extension Center, 115
Northeast Missouri Grain LLC, 122
Northside Planting LLC, 104
no-till, 41, 90, 93, 94, 150, 159

O

Oak Ridge National Laboratory, 151, 153, 159, 160, 177
octane, 42-45, 53, 54, 66, 71, 72, 74, 129
octane boosters, 42, 43, 44
Ohio, 129
oil, 11-19, 23-29, 31-36, 39, 40, 43, 46, 47, 56, 65, 66, 84, 85, 88, 89, 99, 109, 113, 114, 126-128, 143, 150, 155-157, 158, 167, 169, 170, 175, 176, 177, 180
oil companies, 17, 19, 25, 26
oil consumption, 17, 25, 27
oil imports, 31, 34, 35, 175
oil prices, 11, 16, 17-19, 23, 26, 28, 47, 85, 176
oil production, 25, 26, 28, 29, 34, 36, 114, 177, 180
oil reserves, 24-26, 34, 46, 121
oil sourcing, 46
oil well, 12, 24
Oklahoma State University, 106
Öko-Instituts Darmstadt, 147
on-farm research, 41, 90
Ontario, 54, 63, 107, 156
OPEC, 17
organic matter, 89, 90, 93, 94, 150
Otter Creek Ethanol, 123
Otto, Nikolas, 12
oxygenate, 18, 43, 44, 51
ozone, 45

P

Panda Ethanol Inc., 125, 126
paper, 119, 137, 148, 153, 155, 156
particulate matter, 45, 121
perennial crops, 41, 46, 96, 154
Persian Gulf, 35
pesticides, 16, 46, 84, 88-90, 96, 168
petroleum, 11-19, 23-29, 31-36, 39, 40, 43, 46, 47, 56, 65, 66, 84, 85, 88, 89, 99, 109, 113, 114, 126-128, 143, 150, 155-158, 167, 169, 170, 175-177, 180
Petroleum Replacement Ratio, 166-170
phase separation, 52, 63, 77, 78
phenolic polymer, 138
photosynthesis, 24
pipeline, 36, 127-129, 146, 157, 175
plowing, 150
plug-in hybrid, 72, 73, 79
plug-in trybrid, 79
Poet LLC, 112, 141, 149
pollution, 42, 55, 121, 166
polymers, 137, 138
polysaccharides, 138
poplar trees, 41, 151, 153
power alcohol, 13, 175
Practical Farmers of Iowa, 90, 91
precision farming, 41, 90, 172
pre-detonation, 13, 42, 53, 71, 170
Primenergy LLC, 123
Process Design Center, 77, 79, 118, 171,
process fuels, 39, 99, 105, 114-117, 119, 121-123, 125, 126, 168
production cost, 26, 53, 59, 78, 100, 104, 118, 129
prohibition, 11, 13, 14
proof, 76, 110, 127
Prosperity Beckons, 14
protein, 85, 86, 88, 107, 109, 110, 113, 115, 149, 157
Purdue University, 86, 117, 153
pyrolysis, 155, 157

Q

Quadricycle, 12
Quebec, 106

R

Radford University, 13
rain forest, 42
Range Fuels Inc., 142
raw starch hydrolysis, 117, 118
reed canarygrass, 41
refineries, 36
regional biomass processing centers, 156, 157
Reinach, Dr. Fernando, 42, 77, 103, 149, 154
Renewable Agricultural Energy Inc., 110, 149
Renewable Fuels Association, 19, 31, 63, 107
residue harvest, 150
rice straw, 142, 150

S

Saab BioPower, 69, 71, 72
saccharification, 117, 144, 157
Saudi Arabia, 34
Schechinger, Tom, 157
Schlicher, Dr. Martha, 110, 119, 149
Schmidt, Lanny, 76
Sebesta Blomberg, 123
security, 32, 59, 62, 65, 84, 87, 121, 167
service stations, 14, 17, 58, 74, 77, 147
Simmons, Matthew, 26, 28
soil carbon, 46, 94, 96, 150
soil erosion, 41, 88-91, 93-95, 106, 150
soil fertility, 41, 96
soil structure, 93, 94
solar energy, 24, 151, 167, 179
solid waste, 101, 119-121
Sorganol Production Co. Inc., 105
Souixland Ethanol LLC, 120
Soviet Union, 25
soybeans, 94, 115, 116, 151, 159
Spain, 140
Spieldoch, Alexandra, 87
spills, 43, 47, 55
starch, 46, 85, 87, 99, 100, 102, 103, 107, 109, 112, 115-117, 148, 177
starch feedstocks, 99, 117
stillage, 107
storage tanks, 58
Strait of Hormuz, 35
stubble, 93, 139, 150
subsidies, 32, 34, 83, 84, 90, 91
sugar, 42, 99-101, 103-108, 117, 126, 128, 137-139, 141, 144, 145, 148-151, 153, 154, 175, 177, 180
sugar beets, 99, 101, 103, 108, 126, 180
sugar cane, 42, 99, 101-106, 108, 126, 137, 149-151, 153, 154, 180
sugar feedstocks, 103
Sustainable Agriculture Research and Education, 108
sustainable farming, 41, 84, 89, 90, 93, 104, 175
sweet potato, 102, 107
sweet sorghum, 99, 101-103, 105, 126, 129, 151, 154, 180
switchgrass, 41, 46, 47, 96, 139, 142, 151, 152, 155, 156

synfuels, 95
Syntec Biofuel Research Inc., 121
synthesis gas, 99, 119, 139, 140, 143, 145, 146, 158
synthetic rubber, 15, 16

T

tanker, 46
tar sands, 25
taxes, 11, 12, 17, 26, 31-34, 53, 58, 118, 127, 158
terrorism, 35
Thomas, Dr. Sandy, 76
Thompson, Dick and Sharon, 90
tobacco, 107
torque, 69, 71, 72, 75
toxicity, 43, 55
tractor fuel, 16, 46
triticale, 93
Tschaplinski, Tim, 153
turbocharging, 71, 72, 75
twenty-first amendment, 13

U

U.S. Army, 138, 160, 177
U.S. Department of Agriculture, 34, 41, 46, 83-86, 91-94, 96, 102-104, 108, 115, 116, 125, 139, 149, 150, 153-155, 167
U.S. Department of Energy, 36, 45, 52, 55, 56, 58-60, 63, 69, 71, 93, 111, 114, 137, 139, 141, 144-146, 148-150, 158, 160, 175, 178-180
U.S. Department of Energy Biomass Program, 137, 144, 145, 160
U.S. DOE National Renewable Energy Laboratory, 63, 93, 141, 146
U.S. Energy Information Administration, 23, 25, 27, 29, 172
U.S. Energy Partners LLC, 122
U.S. Environmental Protection Agency, 43, 52, 53, 55, 59-61, 66, 67, 69, 120, 122, 124, 129, 147, 159, 170, 171, 179
U.S. Federal Trade Commission, 26, 29
U.S. Geological Survey, 25
U.S. Government Accountability Office, 29

U.S. Government Services Administration, 58
U.S. House Committee on Ways and Means, 87
U.S. National Renewable Energy Laboratory, 128
U.S. Senate Foreign Relations Committee, 35
U.S. Tobacco Tax and Trade Bureau, 127
unconventional reserves, 25
United State Army, 29
University of Alberta, 157
University of California, 102, 107
University of Minnesota, 76, 95
upgraded biogas, 146, 147
USDA Ecomomic Research Service, 83-86, 92
used cars, 68

V

variables, 171
Vehicle Identification Number, 57
VeraSun Energy Corporation, 113
Virginia Polytechnic Institute, 116
volatility, 44, 45, 56, 119
Votarim New Business Ventures, 149

W

waste beverage, 101
wastewater, 118, 126
water use, 86, 118, 119
watermelons, 107
Weimer, Paul, 154
West, Daniel, 108
wet distiller's grain, 117, 125
wet milling, 86, 109
wheat, 46, 93, 101, 102, 115, 116, 139, 142, 150, 159, 180
wheat straw, 139, 142, 150
wildlife, 96, 119
Williams Company, 127
windfall fruit, 101
wood, 37, 119, 121, 123, 140, 141, 143, 144, 146, 148, 153-155, 158
wood chips, 121, 158
wood glue, 154
world hunger, 86

World War I, 13-15, 89

X

XL Dairy Group Inc., 114, 126

Y

yeast fermentation, 109
yields, 14, 42, 86, 89-91, 93, 95, 102, 103, 105-108, 110, 115, 151-153, 158, 15

Printed in the United States
117920LV00003B/281/A